What Should Be Classified?

A FRAMEWORK WITH APPLICATION TO THE GLOBAL FORCE MANAGEMENT DATA INITIATIVE

Martin C. Libicki | Brian A. Jackson
David R. Frelinger | Beth E. Lachman
Cesse Ip | Nidhi Kalra

Prepared for the Joint Staff J-8

Approved for public release; distribution unlimited

NATIONAL DEFENSE RESEARCH INSTITUTE

This research was prepared for the Joint Staff Director for Force Structure, Resource, and Assessment (J-8) and conducted within the Acquisition and Technology Policy Center of the RAND National Defense Research Institute, a federally funded research and development center sponsored by the Office of the Secretary of Defense, the Joint Staff, the Unified Combatant Commands, the Navy, the Marine Corps, the defense agencies, and the defense Intelligence Community under Contract W74V8H-06-C-0002.

Library of Congress Control Number: 2010940485

ISBN: 978-0-8330-5001-4

Published 2010 by the RAND Corporation
1776 Main Street, P.O. Box 2138, Santa Monica, CA 90407-2138
1200 South Hayes Street, Arlington, VA 22202-5050
4570 Fifth Avenue, Suite 600, Pittsburgh, PA 15213-2665
RAND URL: http://www.rand.org/
To order RAND documents or to obtain additional information, contact
Distribution Services: Telephone: (310) 451-7002;
Fax: (310) 451-6915; Email: order@rand.org

Preface

Many Department of Defense (DoD) organizations are developing large-scale integrated data systems that bring together databases from multiple sources and for multiple users through DoD networks. The Global Force Management Data Initiative (GFM DI) is one such system for sharing DoD-authorized force-structure information. GFM DI makes the entire DoD-authorized force structure visible, understandable, and accessible in a common format for the first time, which can help support a wide range of DoD business, readiness, and force management systems.

GFM DI offers DoD users many potential benefits, among them a complete picture, derived from unclassified sources and systems, of the force structure. If that picture is not protected appropriately, however, adversaries might also benefit in that gaining access to it would offer significant advantages over the usual practice of gathering data piecemeal. Thus, analyzing the security vulnerabilities of and potential security mitigation approaches for force-structure data in GFM DI is essential.

For this assessment, RAND Corporation researchers developed a general framework for judging classification decisions and then analyzed the material GFM DI brings together to determine whether it met such criteria. From this assessment, we developed recommendations about how to handle potential vulnerabilities associated with GFM DI. This monograph documents this assessment and recommendations.

This monograph should be of interest to those involved or interested in developing and using GFM DI and other DoD large-scale data systems. It should also be of interest to other government policymakers and managers who are interested in when information, especially when brought together from different data sources, needs to be classified.

This research was sponsored by the Joint Staff Director for Force Structure, Resource, and Assessment (J-8), Models and Analysis Support Office, and was funded by the DoD Modeling and Simulation Office (under the "Vulnerability Assessment for Compilation of Data" project) and conducted within the Acquisition and Technology Policy Center of the RAND National Defense Research Institute, a federally funded research and development center sponsored by the Office of the Secretary of Defense, the Joint Staff, the Unified Combatant Commands, the Navy, the Marine Corps, the defense agencies, and the defense Intelligence Community.

For more information on RAND's Acquisition and Technology Policy Center, see http://www.rand.org/nsrd/about/atp.html or contact the director (contact information is provided on the web page).

Contents

Figures

Summary

DoD frequently needs detailed force structure data to conduct operational planning and budget programming. Each service maintains these data but in different formats, stored on different systems established at different times, and with service-unique labeling that makes aggregating information across the military services difficult, time consuming, and error prone. To address this problem, DoD has initiated GFM DI.[1] Its objective is to standardize formats and protocols so as to ease the process of collecting and aggregating service data for department-level analysis. The great bulk of the data involved are unclassified and therefore accessible by means of DoD's Nonsecure Internet Protocol Router Network (NIPRNET), an unclassified system. Such systems lack security features (e.g., routine content encryption, virtual air-gapping) of the Secret Internet Protocol Router Network (SIPRNET). Because potential adversaries have repeatedly taken data from the NIPRnet, which, broadly speaking, is Internet-accessible, the question has been raised as to whether detailed force-structure data should be stored on SIPRNET exclusively.

The Joint Staff asked NDRI to examine this issue and make recommendations about the need to classify GFM DI information. We addressed this question in two steps. The first was to analyze the reasons for classifying information in general with an eye to distilling some broad criteria that could guide subsequent analysis. The second was to apply the criteria to the data covered by GFM DI.

---

[1]   GFM DI is not a database but a set of standards and connectivity protocols that facilitates the sharing of information stored in various virtual and physical locations.

Why Classify Information?

The most important reason to classify information (and the reason most relevant to GFM DI) is the belief that, if potential adversaries get hold of it, they can use it to undermine U.S. national security. Guidelines exist for determining what national security information should be protected and at what classification level in a tiered system. Assignment to a given level depends on an assessment of how much damage would result if an adversary got the information (e.g., "serious damage" or "exceptionally grave damage").

Some classification decisions—e.g., protecting the identity of a covert intelligence source—are straightforward and prompt little controversy. But other cases are less clear. When does the loss of control over a piece of information lead to "serious damage" or, worse, "exceptionally grave damage"? Without a clear answer, the default decision may be that the information is classified, perhaps overclassified. Classification imposes costs, and these are not just administrative. For one thing, it makes doing business within or among government agencies more difficult. For another, the public has less information about its government.

The "precautionary approach" simply assumes that secrecy confers a benefit and therefore ignores such costs. This avoidance, however, is clearly inadequate for policymaking, which normally requires comparing costs and benefits—understood to include factors that cannot be monetized—to one another.

Thus, to get a handle on the potential damage from adversary gaining specific information, we generated a set of four basic criteria for assessing whether the classification of a particular piece of information has any value:

1. Does classification decrease the amount of information going to potential state and nonstate adversaries?
2. Does the additional information adversaries would have if it is not classified affect what adversaries know (and are such changes meaningful and helpful in the sense that the additional

information moves them closer to, rather than farther from, the truth)?

3. How likely is this change in knowledge to affect possible adversary decisions (and again, does it do so in ways that help the adversary)?

4. Would the decisions the adversary makes based on such knowledge damage U.S. national security?

Only if the failure to classify a piece of information means that an adversary is more likely to get it *and* if having it changes the adversary's estimate of a key piece of knowledge *and* if the change in knowledge alters a decision (or the probability of a decision) *and* if this decision is adverse to the United States would any case exist for classifying it—and then only if the costs of classification, broadly understood, are not greater. If classification yields no measurable benefit, there is no justification for it even if the *costs* of classification are zero, which they never are. In principle, knowledge is power, but not all knowledge is equally powerful.

GFM DI: What It Is, and What Is New About It

GFM DI deals with force-structure information: types of military units and the people who staff them. More specifically, it deals with authorized forces, those detailed in military authorization documents. These forces differ in number from on-hand forces, which are typically those Congress has agreed to pay for, and ready forces, which are those that can actually deploy at any given moment. In the current U.S. wartime posture, both categories are typically smaller than authorized forces.

As mentioned, GFM DI is not a database and does not necessarily require creation of new databases. However, it does mandate that the military services provide a minimum set of essential data about their force elements. The GFM DI data dictionary defines the legitimate values for these required data elements. GFM DI helps integrate service-authorized force management data by allowing users to access what used to be scattered heterogeneous information sources as if they were one coherent database. However, from the classification standpoint, it is *not* a single database, and dealing with security concerns requires identifying security issues related to either the broader sharing of data or to the aggregation of different data types as a result of the initiative.

What Security Risks Does GFM DI Pose?

The security concerns raised about GFM DI rest on the potential for adversaries—states and nonstate actors alike—to use its data. We framed these concerns in terms of the following three questions:

- Will GFM DI provide adversaries information about the U.S. force structure that they would not otherwise have?
- Will GFM DI make it easier for adversaries to confirm information that they may know already about U.S. forces?
- Will GFM DI's aggregation capabilities create security concerns?

For each question, we examined how the changes that GFM DI would require or possibly induce would affect the access potential adversaries might have to such data. We then examined the data and asked, for each concern, whether the classification or other restrictions on it were supported, based on our four basic criteria.

Having laid out a systematic process for examining those security concerns and determined what the security benefits would be from classification, we analyzed the minimum data set and **found no good reason to classify GFM DI as a whole.** Concerns about the standardization, mandated generation of the minimum data set, and broader utilization of force-structure data appear largely unfounded. In considering how much of an overall picture of U.S. force structure adversaries might gain, we noted that the change was from a status quo of many alternative sources of information to slightly better data, a limited decisionmaking advantage. In the end, the concerns generally failed at least one criterion for considering classification.

Two concerns cannot be dismissed, however. First, within the requirements GFM DI imposes on data providers—that they conform to the GFM DI data dictionary and provide the minimum data set might be revealed about a sensitive unit or platform based on the characteristics of the billets associated with it. Linking GFM DI data to personnel databases (external to GFM DI)—the linkage of *individuals* to billets or units—may also create problems.

Data providers have the flexibility to make data beyond the minimum data set available to GFM DI subscribers. Depending on what data are added to service data sets, adversaries might get access to sensitive data not readily available elsewhere. The main example of this possibility was the use of additional fields to add information about individual force structure components. Although flexibility in building force structure databases and customizing them as needs evolve may benefit DoD, the new data they support—whose extent and nature cannot be predicted now—may create security concerns.

Such concerns are not unique to GFM DI (e.g., sensitive information can be inadvertently released in unclassified email). As a result, no good case could be made that their existence mandates that significant parts of the data covered by GFM DI be classified. The potential security concerns raised can be addressed with better tailored recommendations.

Recommendations

First, caution must be exercised when creating additional data fields or when adding data beyond the minimum data set. Someone (e.g., on or designated by the Joint Staff or the Office of the Secretary of Defense) should periodically scan GFM DI to look for information that should not be there.

Second, the list of displayed fields and the allowable attributes for certain data elements may need to be trimmed. For instance, listing the required security classification of billets may reveal information about individuals that may make them targets for recruitment by adversary intelligence organizations. Similarly, some military occupational codes are inherently sensitive; a unit with an unexpected number of such billets calls attention to itself as having unexpected missions.

Third, information on units, platforms, or activities that now guard their security through obscurity may have to be classified. Manipulating data is getting easier over time, and people routinely disclose information in a myriad of intentional and unintentional ways, making things simpler for an opponent. The half-life of such tactics is short—with or without GFM DI—and prudent planners should anticipate as much and adjust accordingly.

Fourth, the interaction between GFM DI and personnel databases (external to GFM DI) needs further study. Technically, this issue is outside the purview of GFM DI developers but should, nevertheless, be examined. The ability to link a person to a billet and a billet to a unit may reveal a great deal more about the unit than would GFM DI data alone. The ability to link persons to each other (by linking person to billet to unit to billet to person) allows potential adversaries to conduct a great deal of social network analysis.

Acknowledgments

The research team would like to gratefully acknowledge the support and productive technical conversations with our research sponsors at the Joint Staff J-8 Models and Analysis Support Office, including George Sprung, Chief, Models and Analysis Support Office; CDR Andrew Thaeler, Sam Chamberlain; LTC Glenn Hillis; and Doug Lamb.

We would also like to acknowledge the input received from the Interagency Operational Security Support. Our interactions with the staff as we refined our ideas and analysis were very useful.

The manuscript profited greatly from the comments and recommendations of our reviewers Robert Anderson and Kenneth Horn, both of RAND. We also acknowledge Jerry Sollinger's help. Any errors of fact and judgment that remain are solely those of the authors.

Abbreviations

ANG	Air National Guard
ARNG	Army National Guard
BFT	blue force tracking
C2	command and control
C3I	command, control, communications, and intelligence
DMDC	Defense Manpower Data Center
DoD	U.S. Department of Defense
DRRS	Defense Readiness Reporting System
DRRS-N	Defense Readiness Reporting System–Navy
eJMAPS	Electronic Joint Manpower and Personnel System
FM	field manual
FMID	force module identifier
FMS	force module subsystem
GCSS-AF	Global Combat Support System–Air Force
GFM DI	Global Force Management Data Initiative
GVC	global visibility capability
J-8	Director for Force Structure, Resource, and Assessment, Joint Staff
MASO	Models and Analysis Support Office

MOS	military occupational specialty
NIPRNET	Nonsecure Internet Protocol Router Network
NDRI	National Defense Research Institute
NOS	not otherwise specified
NSA	National Security Agency
org server	organization server
OSD	Office of the Secretary of Defense
OSINT	open-source intelligence
OUID	organization-unique identifier
POM	program objective memorandum
PPBE	planning, programming, budgeting, and execution
RMS	root-mean-square
SCI	sensitive compartmented information
SIPRNET	Secret Internet Protocol Router Network
TFSMS	Total Force Structure Management System
TRADOC	U.S. Training and Doctrine Command
USA	U.S. Army
USAF	U.S. Air Force
USMC	U.S. Marine Corps
USN	U.S. Navy
XML	Extensible Markup Language

CHAPTER ONE

Introduction

Why classify information? Governments have several reasons for doing so. The information in question could embarrass someone. It could have been supplied by others—ranging from foreign countries to human sources—who expect the government to keep it secret. It could be classified not because it is sensitive per se but because *how and where the government obtained it* is sensitive. That is so because its very existence implies its source, and an adversary (or even a friend) who discovers such sources or the related methods may take steps to eliminate them or degrade their value. In the main, however, the most important reason to classify information is the belief that, if it falls into the hands of adversaries, they can use it undermine national security.[1]

The United States has guidelines for what national security information should be protected and by how much classification. The terms *damage*, *serious damage*, and *exceptionally grave damage* are used to determine whether a piece of information falls into the categories of Confidential, Secret, and Top Secret, respectively.[2] But beyond these, neither executive orders nor Department of Defense (DoD) instruc-

---

[1] For a fuller explanation of classification reasons and rationales, see discussion in Arvin S. Quist, *Security Classification of Information*, Vol. 2: *Principles for Classification of Information*, April 1993. We had initially thought that a literature search would turn up numerous references of previous research on criteria for classifying information. However, other than Quist's work, such references are not available.

[2] See the discussion of classification levels and their definitions in Executive Order 12958, "Classified National Security Information," as amended through March 28, 2003. See also Assistant Secretary of Defense for Command, Control, Communications, and Intelligence, *Information Security Program*, Washington, D.C., DoD Regulation 5200.1, January 1997.

tions provide any guidance on how to determine whether or when such information can cause how much damage.

In some cases, the damage that losing control over a piece of information can cause is obvious; for example, revealing the identity of a confidential intelligence source would not only harm the source and forfeit any future information that source might have provided but would also make recruiting other sources far more difficult. Choosing to classify information in such cases is therefore straightforward and creates little disagreement or controversy.

But in other cases, calibrating the scope of the "damage" and what constitutes "serious damage" or "exceptionally grave damage" is anything but obvious. This ambiguity complicates decisions about what information should be classified. The result often defaults to broad classification out of an abundance of caution—if the classifier can envision any way an adversary might use a piece of data, classification might appear warranted by applying the security equivalent of the precautionary principle. But caution risks overclassification, which carries costs for both the public (which is denied information about what the government is doing and how well it is doing it[3]) and the function-

---

Note that classification is a means for which restriction is the end. The distinction between the two will be further elaborated in this and the next chapter.

[3] See Quist, 1993, particularly Ch. 5, "Information Disclosure Risks and Benefits," for a discussion of how information disclosure can have associated benefits for national security in addition to posing risks. Such concerns have existed for many years with respect to national security and protection of Secret material (see William E. Colby, "Intelligence Secrecy and Security in a Free Society," *International Security*, Vol. 1, No. 2, Autumn 1976. In early policy statements on classification and information release, the Obama administration emphasized disclosure over restriction:

> including the possible restoration of the presumption against classification, which would preclude classification of information where there is significant doubt about the need for such classification, and the implementation of increased accountability for classification decisions. (Barack Obama, "Classified Information and Controlled Unclassified Information," memorandum for the heads of executive departments and agencies, May 27, 2009)

See also the administration memorandums and policy statements Steven Aftergood has posted on the Federation of American Scientists' website.

ing of government itself.[4] It has monetary costs as well.[5] Classification hampers the ability to share information with people who need it
(e.g., state and local officials, private firms) to do their jobs (counterter-
rorism). Some restrictions also can constrain security efforts otherwise
advanced by disclosure. The price may be worth paying, but, at very
least, the value of what it buys must be clear.

Many classification decisions are based on regulation, precedent,
and analogy—as well as on speculation, howsoever informed. If information looks like something that an adversary's intelligence services
might have an interest in, it is easy to conclude that its circulation
should be restricted—whether or not the classifier can articulate how
the national security would be damaged should others get the data.
Some precedents are so institutionalized that classification decisions
are based largely on who owns or created the information. Such restrictions, as with much government regulation (e.g., military performance
specifications), may be no better than the fossilized memory of ear-
lier mistakes—a catalog of things gone wrong that no one wants to

---

[4] In the words of the Information Security Oversight Office (ISOO):

> Classification, of course, can be a double-edged sword. Limitations on dissemination of
> information that are designed to deny information to the enemy on the battlefield can
> increase the risk that our own forces will be unaware of important information, con-
> tributing to the potential for friendly fire incidents or other failures. Likewise, imposing
> strict compartmentalization of information obtained from human agents increases the
> risk that a government official with access to other information that could cast doubt
> on the reliability of the agent would not know of the use of that agent's information
> elsewhere in the Government. The National Commission on Terrorist Attacks Upon the
> United States noted that while it could not state for certain that the sharing of informa-
> tion would have succeeded in disrupting the 9/11 plot, it could state that the failure to
> share information contributed to the government's failure to interrupt the plot. Simply
> put, secrecy comes at a price. For classification to work, agency officials must become
> more successful in factoring this reality into the overall risk equation when making clas-
> sification decisions. (ISOO, *Report to the President 2004*, Washington, D.C.: National
> Archives and Records Administration, March 31, 2005)

[5] See discussion in ISOO, Cost Report for Fiscal Year 2008, Washington, D.C.: National
Archives and Records Administration, May 19, 2009b. The report includes an annual estimate of $8.64 billion, not including the expenditures of the Central Intelligence Agency,
Defense Intelligence Agency, National Geospatial Intelligence Agency, National Reconnaissance Office, and National Security Agency. These organizations consider their budgets to
be classified information (ISOO, 2009b, p. 2).

come anywhere close to repeating. There are times when precedent and analogy are effective heuristics. They are consistent with how humans usually deal with new situations, and in many situations, they are the best way to think.[6] People may be hardwired to use such methods. However, mechanistic use of classification criteria can overwhelm good sense.[7]

Hence, it is necessary to examine the conceptual foundations for decisions normally made on the basis of precedent and analogy—if only because what works well in one era works less well in another. Specifically, the end of the Cold War, coupled with the advent of ubiquitous Internet computing, requires rethinking how information is distributed.[8] The era of information scarcity is rapidly fading.[9]

Classification and the Global Force Management Data Initiative

In examining the issue, we focused on one particular family of data, the information addressed in the DoD Global Force Management Data Initiative (GFM DI). We did so at the request of the Joint Staff to determine whether and to what extent the information that the ini-

---

[6] In Gary Klein, *Sources of Power: How People Make Decisions*, Cambridge, Mass.: MIT Press, 1998.

[7] As can the mechanistic use of the less-formal designation of certain information as "security sensitive."

[8] The end of the Cold War is relevant because it altered the nature of the harm that could result from potential adversaries gaining a piece of information. The Soviet Union could make very consequential and deleterious decisions based on such information. Can the same always be said of today's potential adversaries, which are variously less powerful (e.g., Iran) or less hostile (e.g., Russia compared to the Soviet Union)?

[9] Reinforcing the need for such a revisiting, in its review of federal classification guides—the documents intended to frame classification choices—ISOO found that 67 percent of the guides currently in use had not been updated in the last five years (ISOO, *Report to the President 2008*, Washington, D.C.: National Archives and Records Administration, January 12, 2009a, p. 23).

tiative addresses and that is unclassified should remain unclassified or, instead, be restricted.[10]

GFM DI aims to facilitate sharing of information on the authorized force structure of the U.S. military within DoD. The goal is to make such information more available within DoD, increasing the efficiency, effectiveness, and accuracy of management and other processes that rely on force-management data. The initiative and its goal are discussed in greater detail in Chapter Three. It may be considered an expanded version of what the Army calls a (modified) table of organization and equipment but with specifics on billets, military units, and platforms.

Technically, GFM DI is not itself a database but a set of standards and connectivity protocols that facilitates the sharing of information stored in various virtual and physical locations. GFM DI does not contain information of its own. Furthermore, most of the information it addresses would not be created de novo to comply with its requirements; after all, each service keeps force-structure information today. What GFM DI does is rationalize and standardize the presentation and management of that data across DoD to make it more accessible across and beyond the services and hence more useful. Unfortunately, improving the accessibility of information to authorized users also has the potential to make it more accessible to unfriendly states and hostile nonstate actors.

The fact that the GFM data lack features that would otherwise complicate a discussion of classification made our analysis simpler. The basis for the data is straightforward reporting, so the contents of GFM DI can be evaluated for what they reveal, rather than what they might suggest about how the data was produced. GFM DI is not an intelligence system, where the data itself might reveal something about the sources and methods of its collection or analysis. Furthermore, since the material in GFM DI is restricted to information about and generated by the U.S. military, few if any diplomatic issues confounded our analysis: We found no material that should be classified solely because some foreign country would legitimately consider it sensitive.

---

[10] GFM DI has a classified side, but we did not analyze it.

Hence our question: What are the risks of keeping all or most of what is now unclassified GFM DI information unclassified even after aggregation becomes easier? To examine that question, we focused on whether the contents of GFM DI—in total or in part—should be classified because of what it might reveal to current or potential adversaries or competitors about U.S. military capabilities, intentions, or activities. As subsequent chapters will discuss in greater detail, doing so required addressing three specific issues:

- *What information does GFM DI cover?* The issue here is what individual pieces or aggregations of data and what sensitive information they might reveal about the U.S. military.
- *What was actually changing as a result of implementation of GFM DI?* GFM DI does not directly create new data, so we are really asking how the implementation of GFM DI changes the data that adversaries will be able to collect.
- *What would be the appropriate rationale for concluding that information covered by GFM DI should be classified?* The issue here is weighing concerns about operational security and intelligence.

About This Document

The following chapters present the results of this examination. Chapter Two presents our framework for thinking through and making classification decisions. Although our thinking focused on force-structure information, we have laid out a logic that is more generally relevant to any classification of information. Chapter Three describes GFM DI and discusses what its implementation changes. Chapter Four presents our analysis of the initiative, structured by the security concerns raised about GFM DI and our evaluation of them, based on the framework laid out in Chapter Two. Chapter Five concludes the discussion and presents the findings of the analysis.

A Framework for Classification Decisions

Information is classified to ensure that it does not end up in the hands of adversary states or others in ways that would hurt the national security interest.[1] Although governments and their component organizations sometimes have other reasons and incentives to restrict information and may use security classification systems to do so, the fundamental reason for classification—in the United States at least—is to protect security interests.[2]

Apart from situations in which the security value of classification is obvious—e.g., protecting the identity of a clandestine source—how should decisions be made about what pieces of data should be classified? Since classifying information creates costs, it should be approached as an explicit cost-benefit comparison (understood to include factors that cannot be monetized).[3] Such a comparison necessarily goes beyond a "precautionary approach"—one that assumes that there are benefits to being cautious and applying secrecy irrespective of the costs of a

---

[1] In considering adversaries or competitors, classification is often thought of in terms of enemy governments, but nations may have an interest in concealing information from governments of nations with which they are not in active military conflict (and even governments they are never likely to fight). Even friendly and allied governments have things they hide from each other. Governments have an interest in keeping such information as details of operations or vulnerabilities of potential targets from nonstate actors—notably, terrorists. Other important nonstate actors include criminal enterprises.

[2] See Daniel P. Monynihan, *Secrecy: The American Experience*, New Haven, Conn.: Yale University Press, 1998, for an excellent discussion.

[3] See Quist, 1993.

classification regime.[4] Simply applying the precautionary principle is insufficient where the benefit from the broad availability and ability to share information among friendly forces may outweigh the risk that an enemy may use that information for its own ends.[5]

In framing a benefit-cost comparison, we would ideally start with discussion of the rationales behind past classification. Unfortunately for analysis, such processes and the results of applying them often go unrecorded or are themselves classified because of what they might reveal.

Thus, instead, we built a framework to elaborate the fundamental rationale behind classification—that there should be security benefits from applying it. Making such a comparison is admittedly difficult, relying both on what we know about adversaries and on how they use information and seeking to anticipate how particular pieces of information might or might not be used in the future. For real world data, the examination becomes complex quickly. As a result, in our analysis, we did two things to provide a basis for our look at GFM DI.

We examined a set of abstract examples of how adversaries might use information. Assigning heuristic numbers to the different parts of the decision illustrated how statistical decision theory could be applied to the analysis to permit an integrated comparison of costs and benefits (broadly understood). Because we focused on GFM DI, these examples look at force-structure (and associated billet and unit) information. Our analysis, particularly in Chapters Three and Four, also reflects the facts that GFM DI is more like a portal to information and that the information itself tends to change slowly.

---

[4] The tendency to such a precautionary approach is reinforced by the fact that, in general, those making the classification decision will not bear the full costs (or potentially any of the costs) of restricting the information.

[5] Executive Order 12958 states that classification is warranted *only* if "unauthorized disclosure of the information reasonably could be expected to result in damage to the national security" and the "original classification authority is able to identify or describe the damage." It does not necessarily mandate classification even if these conditions are met. Thus, nothing in the executive order precludes making a comparison of costs and benefits in deciding whether to classify a piece of information.

We endeavored to remain *practical*. It is well and good to speak of balancing costs and benefits in decisionmaking, but costs and benefits that are hard to measure leave little for analysts to work with. Our goal was to produce an understandable set of criteria for considering classification choices. These criteria—based on the presumption that classification must be merited by the security benefits—are laid out in the remainder of this chapter.

Considering the Benefits of Classification

Traditional classification assessment focuses on the benefits of classification in terms of how much adversaries appear to value what they can steal from us. This value, in turn, translates directly into a "degree of damage" associated with the information's disclosure or, put another way, the degree of benefit potentially associated with the information's concealment.[6] All this then colors how much to restrict information, that is, what rules govern dealing with it. These vary from the stringent restrictions associated with the most highly classified material to the simpler procedural restrictions around official use–type data—or whether it can be used more freely and with fewer safeguards on its storage, transfer, copying, and destruction.[7]

The public debate about classification policy does raise the question of whether the degree of damage is being estimated well. Put simply, just because a specific piece of information or a data set is

---

[6] See, for example, the discussion in Quist, 1993, Ch. 5.

[7] Classification is a technique that gives information legal protection and that triggers a coherent set of information security policies. These policies include, relevantly, computer security, e.g., classified information cannot be stored on unclassified systems, which, in turn, are the only ones accessible from the Internet. They also include physical security (e.g., paragraph markings, document covers, document storage, where they can be displayed, and how they are transported) and personnel security (users must be cleared to the security level of the material and, in some cases, must demonstrate need to know). One could easily imagine protection mechanisms other than classification, such as the use of end-to-end encryption, strong accountability in computer use, user-level watermarking (to detect where information leaked from), and deception (which would be transparent to legitimate users and completely obfuscating to others).

useful in some way and relates to areas of security concern, it does not necessarily follow that the same information is useful to an adversary. Indeed, knowing that potential adversaries are interested in the information is no proof that their satisfaction would damage U.S. national security. If it is not damaging, restricting access to it will not, in fact, produce the expected security benefit.

Although the nature of secrecy itself makes it difficult to evaluate many specific classification decisions, there are cases of formerly open data being classified retroactively. After September 11, 2001, for instance, government agencies became concerned that terrorists could exploit open geospatial databases on national infrastructures to help plan future attacks. A RAND Corporation study of such databases concluded that, in many cases, information we found quite useful (and that was so because it was widely shared and usable) was of very limited utility to terrorist groups because they simply did not need such information to plan attacks.[8] The broader the precautionary approach taken to information restriction, the more likely it is that data will be restricted when, in fact, restriction offers little to no security benefit.

Thus, when considering whether a piece of information should be classified, it is necessary to go beyond the precautionary principle and to fundamental questions. The classification decision should be treated like any other decision, comparing its costs and benefits (as broadly understood): If the government does not classify this data, what would the potential damage to national security be?

We drew from statistical decision theory to frame four basic criteria for assessing the value of classifying a particular piece of information:[9]

1. Classification must decrease the amount of information going from the United States to potential state and nonstate adversaries.

---

[8] See John C. Baker, Beth E. Lachman, David R. Frelinger, Kevin M. O'Connell, Alexander C. Hou, Michael S. Tseng, David Orletsky, and Charles Yost, *Mapping the Risks: Assessing the Homeland Security Implications of Publicly Available Geospatial Information*, Santa Monica, Calif.: RAND Corporation, MG-142-NGA, 2004.

[9] John W. Pratt, Howard Raiffa, and Robert Schlaifer, *Introduction to Statistical Decision Theory*, Cambridge, Mass.: MIT Press, 1995.

2. The information in question must affect the overall knowledge that such adversaries have in putatively nontrivial and helpful ways; thus, having the knowledge would move the adversary closer to, rather than further from, the truth.

3. The change in overall knowledge must be of the sort that would plausibly affect decisions that the adversary makes or might make in nontrivial or helpful ways.[10]

4. The decisions the adversary would make could damage U.S. national security.

If the failure to classify a piece of information means that an adversary is more likely to have it, if having it changes the adversary's estimate of a key piece of knowledge, if the change in knowledge alters a decision (or the probability of a decision), *and* if this decision is adverse to the United States, there is a prima facie case for classifying the information—*if* the costs of classifying it are not greater than the costs of not classifying it.

Conversely, if the classification does not change what the adversary learns, if what the adversary learns does not change what it knows (about one or another parameter), if what the adversary knows does not affect what it decides, *or* if its decision is not adverse to U.S. interests, there is *no plausible case* for classifying the information. If there is no measurable benefit from classification, there is no justification for classification, even if the *costs* of classification are zero, which they never are.

These criteria rest on the systematic application of common sense. An adversary cannot damage the United States except by taking

---

[10] It is possible for the knowledge a potential adversary acquires to influence decisions that *others* make—to the detriment of the United States. One adversary may, for instance, take a piece of information and give it to nonstate actors, who then use it against the United States. The adversary may instead release the information publicly in ways that show the United States in a bad light and cause people to withhold their support from the United States or dedicate more energy to opposing the United States. The relevant decisionmakers may be different from the initial knowledge acquirers. The last formulation, however, raises the question of whether certain forms of information should count in an assessment of classification, given the U.S. commitment to the principles of the First Amendment. As it is, GFM DI data rarely, if ever, are of such a nature as to influence third-party views of the United States.

action (or, alternatively not taken)—and not all actions a state or a nonstate actor takes necessarily damage this country. Hence, the third and fourth criteria. If information is to alter decisions, it has to affect the input to these decisions. Assuming reasoning decisionmakers, the information affects decisions by changing what they know to be true (or the confidence with which they know it); otherwise, it never enters the decisionmaking calculus (hence, the second criterion.)[11] Finally, the notion that classification has to change the flow of information that directly or indirectly reaches potential adversaries (or, more broadly, those who can make decisions that would damage U.S. national security) is straightforward. Otherwise, what was the point (hence, the first criterion)?

Two further points need clarification in thinking through these four criteria. It is not sufficient that each criterion can be passed by *some* adversary; at least one adversary that can pass all four criteria has to be identified. To illustrate this distinction, consider the following hypothetical example. Assume there is a piece of information about the layout of a military base in the United States. This information shows a vulnerability that will permit it to be successfully and consequentially attacked by mortars. It is currently held within DoD and is not publicly available. Should it nevertheless be classified? We posit two adversaries: the terrorist and the large state (large in the sense of having access to top-flight hackers, e.g., Russian or Chinese). We assert (1) that classification would reduce the likelihood of the information's distribution, (2) that the data reveal that a base is vulnerable to a particular type of attack, (3) that the revelation of such a vulnerability would make such an attack more likely, and (4) that the consequences of such an attack would damage the United States. Thus, we decide to classify it.

---

[11] Note that we use *reasoning* rather than *rational*—much less *rational* as defined according to some comprehensive cost-benefit standard that an economist might recognize. A decisionmaker may ignore new information altogether, in which case it makes no difference that he or she has it. Alternatively, information may enter his or her decisionmaking in a non-reasoning manner; if so, *knowledge* may be an imprecise term for characterizing how information affected decisionmaking. The basic principle (information affecting decisionmaking) would be the same. Anyway, GFM DI data is not the sort of information that tends to generate an unreasoning response.

But is this a correct decision? Not if we made three more assertions: (1) Although taking this information off the Nonsecure Internet Protocol Router Network (NIPRNET) would keep large states from accessing it, the terrorist, less adept in cyberspace, lacks the skills to access information by breaking into NIPRNET;[12] (2) the large state does not share information with the terrorist; and (3) the large state has no interest in mortaring U.S. bases, perhaps because doing so would raise the risk of general warfare. If so, the information fails our fourfold test for classification when each of the adversaries is viewed individually: The terrorist cannot get the information, even in its unclassified state; the large state cannot use it because it can only inform decisions the state made on other grounds; and the state and the terrorist do not exchange information. Thus, in this hypothetical case, neither threat—the terrorist attack and the large state attack—passes all four criteria. So, classification is unwarranted (although the information should not be released in a form that could allow terrorists to get to it).

The second point is that the results of the four criteria, if probabilistic (and most honest assessments of the effects would be expressed in terms of likelihood) and independent of one another, have to be multiplied together to generate a plausible estimate of how likely damage is.[13] The multiplication of two probabilities almost always produces a

---

[12] Access to the unclassified NIPRNET is limited to authorized users and is currently protected by a variety of mechanisms, not least of which is the requirement to possess a valid Common Access Card and a registered personal identification number to go with it. However, notwithstanding all that, some states have penetrated NIPRNET. For this example, we assumed that nation-states can do so again but that terrorists lack the capability to do so.

[13] Take another hypothetical. For example, the United States has test information on the spread of its Mark XXXVII submunitions that would reveal to a reader that they disperse twice as widely as Mark XXXVI submunitions did. Should it be classified? Investigation reveals that classification would reduce the odds that such information would leak to an active adversary from 15 percent to 5 percent, a 10–percentage point difference. If the adversary had this paper, there is a 10-percent likelihood that it would affect the enemy's assessment of the warhead's effect (e.g., only one circumstance was tested). If the adversary had an altered knowledge of the dispersion pattern, there is a 10-percent chance it would rewrite its training manuals to accommodate the possibility that its ground units would come under a wider attack than they had previously trained for (e.g., most adversary opinion contends that soldiers cannot determine what kind of round is incoming and thus more training would only confuse them). If the training manuals are rewritten, there is only a 10-percent

result that is less probable than each considered separately; if the probabilities are low, cross-multiplying them leads to an event that is far less probable. Admittedly, such calculations presume knowledge of how the adversary would react, something that, in the real world, would never be known with anywhere near requisite precision. So, this is really a test of whether advocates of classification can tell a reasonable story with plausibly nontrivial odds that disclosure would damage national security. We will now discuss each of these criteria in turn.

Denying an Adversary Something

The first criterion is that classifying information must mean that less of it falls into the hands of adversaries. Classifying information is not justified if an adversary cannot get it now, even in its unclassified state or, conversely, if the adversary can get the information despite its being classified. Put another way, classification should not be used unless it is needed.

Here, the question about whether the information covered by GFM DI should be classified is largely about whether the data involved should be moved from the somewhat protected NIPRNET to the more highly protected Secret Internet Protocol Router Network (SIPRNET).[14] Both involve more protections and access controls

---

chance the newly learned reaction of adversary soldiers would affect the performance of the adversary soldiers to the detriment of U.S. soldiers (e.g., the rest of the time, adversary soldiers react such that each individual soldier is somewhat better protected but unit cohesion suffers proportionally). Thus, the odds of harm to U.S. interests are the cross-multiplication of each of four effects—10 percent times 10 percent times 10 percent times 10 percent—or 0.01 percent, multiplied, in turn, by the magnitude of the potential harm. True, these decisions may not be independent. Part of the reason that the odds that the information would affect the training manuals are low may be precisely the possibility that soldiers so trained would protect themselves at the expense of unit cohesion (if the latter is not true, the likelihood of a rewrite rises correspondingly). Nevertheless, if these criteria are largely independent, the magnitude of the resulting harm from not classifying the information has to be at least 10,000 times higher than the harm (if all four conditions are met) from classifying the information to justify classification.

[14] NIPRNET is essentially the "dot mil" portion of the Internet. It has sites that are open to the public (e.g., the Defense Technical Information Center) and many more that are closed. A variety of mechanisms control access to the closed sites. Some do not transfer information to requestors without .mil addresses. Others require specific authorized usernames and pass-

than does information that is freely and publicly available. Moving the information included in GFM DI from NIPRNET to SIPRNET would indeed provide it greater protection from adversaries with the ability to penetrate the former but not the latter. It is important to note, however, that moving it to the higher level of protection would have no actual security benefit against potential adversaries that cannot penetrate NIPRNET —likely the vast majority of nonstate adversaries of concern, such as terrorist groups or criminal enterprises.

Affect Adversary Knowledge

The second criterion is that classifying or restricting information must plausibly reduce what a potential adversary *knows*. There is little point in classifying data that an adversary already knows from other sources. The concept, "know," has two prerequisites: that the information is true and that the adversary has high confidence that the information is true. Both conditions need to apply; after all, it is possible that the adversary has misplaced confidence in a false piece of knowledge—indeed, that is how deception works. This condition is not always fulfilled by public revelation; just because a fact is true and reported in the media does not mean that the adversary credits it completely.

Although it is never certain what potential adversaries do or do not know, it is possible to hazard a good guess using what is freely available (e.g., in various publications, hardcopy or online). We can add broad assumptions about the information collection abilities of specified states or nonstate groups of concerns—e.g., what is likely observable using satellites (whether theirs or a third party's) and what could be learned directly from low-risk human observation in America's open society. The more that alternatives are easier, cheaper, present less risk, and offer sufficient credibility, the less likely it is that acquiring such information from the source under consideration for classification will

---

words, often requiring personal connections. NIPRNET also runs utilities to distinguish, as best as it can, normal ingress-egress traffic from abnormal and potentially troubling traffic and may also have other, less-obvious forms of protection. Although SIPRNET is more heavily protected than NIPRNET, its protections—like those of all information technology systems—are unlikely to be perfect for a variety of reasons. Viruses from the outside, for instance, are not unknown on SIPRNET.

tell adversaries anything they do not already know and, hence, the weaker the argument for further data restriction.[15] Those who argue for maintaining classification for information that has been published must perforce argue that adversaries do not grant such sources sufficient credibility (to act on, as we discuss below).

It might seem that keeping a particular piece of information Secret would reduce an adversary's level of certainty about whatever information they get from other sources. That is not always the case, for a variety of possible reasons, some practical and others driven by the idiosyncrasies of human analysts and decisionmakers. When a piece of information tells someone something that they had no information at all about previously (e.g., that there is a mole operating in their organization), that fact clearly increases their knowledge in an important way. But for other types of information—and for other adversaries—more information does not always necessarily translate to significantly more knowledge.

What an adversary already knows is a significant driver of whether new information will affect an adversary's knowledge. The more the adversary knows before it gains access to the information being considered for classification, the less that the information is likely to change the adversary's understanding. This clearly covers information that directly duplicates what the adversary knows (or should reasonably be expected to know) as a fact. But, in many cases, rather than being identical to existing information, new data sources will enable an adversary to confirm or refine its understanding through comparison of the new and old information. In essence, the new information provides a way to reduce any uncertainty about the existing information; take an example. An opponent's intelligence service might believe that the United States definitely has between 20 and 25 of a particular type of unit and might believe that there is a 40-percent chance that the actual value is 23. Additional information—e.g., intelligence from a new source that

---

[15] For additional discussion of these elements for classification decisionmaking, see Quist, 1993, Ch. 4, "Can the Information Be Controlled by the Government?" The chapter addresses both the questions of whether adversaries already know the information at issue and whether the adversary can readily obtain the information from other sources.

describes U.S. force structure—that indicates that the number is 24 may leave staff members equally certain that the true number is in the 20 to 25 range and an 85-percent chance that the real number is 24. Although this is an improvement (a shift in what they think the true number is and a modest change in the certainty that it falls within a particular range), it is an open question whether that improvement is significant. In many circumstances, it would not be, as we will discuss in more detail below regarding decisionmaking.

What data one can find and what one needs to know to make a decision (see below) are not always the same. One unit, for instance, may want to know where an opposing unit sits. It cannot observe this directly, but it can collect indicators, e.g., what part of the landscape has been disturbed most recently, where noises and lights are coming from, where tanks from that unit have been found, or who among the local inhabitants may have seen soldiers for that unit. None of these says where the unit is, but all are pieces of evidence that suggest where it might be. Specifically, these data shape a notional probability curve over the set of possible locations. Statistical theory indicates that reducing uncertainty bands by a factor of two requires collecting roughly four times as many data points of comparable relevance and quality. Clearly, the last data point received is likely to be only marginally useful compared to the first data point received. More broadly, the more information an adversary already has—assuming that information is not grossly wrong—the less effect an additional data point or data source will have on its level of knowledge. Furthermore, decisionmakers rarely start without initial (a priori) estimates unless the parameter in question is one they simply failed to consider beforehand.

The last example illustrates an important linkage between data (what you get) and knowledge (what you need for a decision). There may be several interdependent steps. For example, to approximate conditions in October 1973, Russian sailors may suddenly be scarce in Cairo markets (data). An observer might infer that they are leaving suddenly (information) and speculate that, since a war would put Russians at risk, their departure means that Egyptians are about to launch an offensive against Israel (knowledge in the sense of an input into a decision).[16] Even if the observational data is completely believed, determining the value of this information requires calculating the likelihood that the observer's belief in the information (that the Russians are leaving suddenly) has been changed by the data (Russians are suddenly scarce in the market) and multiplying it by the likelihood that the belief in the knowledge (the Egyptians are getting ready to invade) has been changed by the information. It is also possible to insert another link in the chain, such as that the Russians have been called back to their boats, which may or may not mean they are leaving—perhaps they are preparing for drills. The longer the chain of inference, the lower the probability, everything else being equal, that the new data are likely to affect the knowledge.

Finally, adversaries are *human* analysts and decisionmakers—with all the strengths, weaknesses, biases, and limitations inherent in that reality. This means that, in some situations, adding additional new information may actually hurt rather than help the state of their knowledge. This situation could obtain, for example, if an adversary used new data selectively, to confirm its existing understandings and biases even when the data do not; if the new data are overweighted for some reason and therefore not interpreted objectively (for example, a source is viewed as more credible than it should be because it is classified or considered official); or if adding more data simply produces information overload, which hurts all decisionmaking. Although whether such things *will* happen as a result of the release of a specific data source will likely be difficult to anticipate before the fact, the possibility that they *could* happen cannot be ignored when considering information security or classification decisions. If, as has been argued, intelligence services are highly susceptible to seeing what they want to see (e.g., weapons of mass destruction in Iraq, statistics on the Soviet gross national product), dissenting reports might disabuse them of their errors, making it important to keep such reports out of adversary hands. Yet, if humans

---

[16] This example draws from Janice Stein, "Calculation, Miscalculation, and Conventional Deterrence: The View from Jerusalem," in Robert Jervis, Richard Ned Lebow, and Janice Gross Stein, *Psychology and Deterrence*, Baltimore, Md.: Johns Hopkins University Press, 1985, p. 74 (the actual details of what happened are different, however).

have an amazing ability to deceive themselves and ignore conflicting data,[17] *none* of this data will change their minds—thus, no damage would befall a nation from leaking such data to adversaries who are determined to ignore it.

Affect Adversary Decisionmaking

Our third criterion is that for classification of information to be justified, it must plausibly affect a decision an adversary is making.

Adversaries or competitors may want to know many things about the United States and its military activities. For their knowledge to be detrimental to the United States, however, it must bear on some decision *on which they are currently undecided* (although they may be leaning in one or another direction). If getting such information has no or a negligibly small effect on what the adversary *does,* it is difficult to articulate how its possession of that data would be negative from the U.S. perspective. The emphasis on *undecided* is critical. If someone is convinced to do something, acquiring new—even clearly relevant—information may have no influence on what is done.

In saying this, it is clear that the concept of a *decision* must be framed broadly enough to encompass circumstances in which receipt of purloined information results in no decision when a presumably less-well-informed (and thus, presumably, ill-advised) decision would otherwise have been made. Information may alter the terms of a decision by removing options (e.g., a choice between A, B, and C now becomes a simpler, less-hazardous choice between A and B). Alternatively, the decision with or without the purloined information would have come out the same anyway but can now be made faster and implemented sooner. Information may also call an adversary's attention to a decision that needs to be made, e.g., some insight into the role of a particular military platform that the adversary had not previously considered which now requires countermeasures.

---

[17] "Defensive avoidance" is one such decisionmaking pathology. See Richard Lebow, "Miscalculation in the South Atlantic, the Origins of the Falklands War," in Jervis, Lebow, and Stein, 1985, pp. 103–107.

To justify classification, the classifier must be able to at least sketch the decisions the information may affect and how gaining access to the information would affect them. Such a sketch needs to take into account how individuals, groups, and larger organizations make decisions. In general, rational individuals base decisions on a number of factors, so for new information to alter the overall outcome of the choice, it must be significant enough to outweigh other elements that are already part of the consideration. When new information reduces uncertainty, how much that reduction will affect many decisionmakers is an open question. Would being 90-percent confident of a particular fact about U.S. forces lead an opposing nation's military planners or operators to a different conclusion than if they were 80 percent confident of that fact? This is not to say that such information will not be sought; people love to get information that confirms previous decisions. But the eagerness for confirmation should not be confused with the possibility that such information would be put to use.

Adverse to Us (and Important)

Our fourth, and potentially most obvious, criterion is that, whatever decision is at issue, the outcome of that decision must matter to the United States. Having the additional data must reasonably lead to choices that damage U.S. national security.[18] Adversaries might glean information on ongoing military actions, future military actions, current capabilities, and future capabilities and might use that information to make decisions that could adversely affect the United States. That damage could be manifested in terms of allowing the adversary to conserve resources, counter U.S. actions, or use forces to exploit U.S. weaknesses.

At first glance, it might seem intuitive that any decision an adversary or competitor would make that is in their interests would automatically be negative for U.S. national security. This would certainly

---

[18] This concept is reflected in the principles guiding classification in government documents on the subject. See, for example, U.S. Department of Energy, *Report of the Fundamental Classification Policy Review Group*, Washington, D.C., October 1997: "Classification must be based on explainable judgments of identifiable risk to national security and no other reason."

be the case between simple adversaries engaged in an all-out war to the death. If both participants in a struggle were committing every resource to the conflict at the expense of all other alternative uses, it would be reasonable to assume that anything that benefitted one (providing an advantage, conserving resources, etc.) by definition damages the other.[19] However, most real-world conflicts cannot be reduced to the caricature of two combatants engaged in an all-out existential struggle. Even the current conflict with al Qaeda does not have this character; decisionmakers on both sides have multiple goals and are devoting resources to multiple ends simultaneously (e.g., both the United States and al Qaeda consider the current Iranian government to be hostile).

In more-realistic conflict situations, better decisions by one opponent may not, in fact, imply an equal cost or disadvantage to the other. As a result, some decisions adversaries make, even if based on better information taken from the United States, may benefit them without costing the United States very much, if anything at all. For example, a confirmed adversary may seek information to inform their diplomatic or military choices about third countries. Such information could even be about U.S. forces, if the nation's ability to respond to their actions is an important part of their calculus. Even though gaining such information might make it easier for them to advance their interests, their resulting decisions (specifically, the effect of the information on their decisions) may have a neutral or even positive effect on U.S. interests. During the Cold War, for instance, it was in the interest of stability— hence in the U.S. interest—that the Soviet Union had some confidence in the true size and disposition of U.S. strategic forces.

Real-world situations are similarly complicated by the fact that the same adversary actions could impose costs on some elements of society while benefiting others—or even the country overall. To use a commercial example, if Chinese hackers, for instance, steal the secrets of a technology that improves their industrial firms' ability to cut their costs and reduces their emissions of greenhouse gasses, this would

---

[19] This assumes that enemy states are unitary actors. Often, they are coalitions of competing bureaucracies that use information stolen from the United States to help them compete with one another for status, prestige, and resources.

most likely be bad news for U.S. companies that compete with them or might otherwise have sold them such equipment. On the other hand, it might be good news for U.S consumers overall and actually advance national goals for addressing global environmental problems.

Stolen information may reduce expenditures in other areas (e.g., knowing that the United States has fewer of a certain system may allow another country to cut its acquisition budget). But whether the adversary's savings is a cost to the United States depends in large part on what the adversary does with the savings. If it uses them to buy more intelligence or more defense, the United States loses. However, to pick an extreme example, if it spent the resources on environmental protection or improving health care for its citizens, the United States might even win.

Considering the Costs of Classification

Traditional operational security analysis pays comparatively little attention to the potential costs of classification. The initial costs include building secure facilities, managing and vetting personnel, and setting up and operating information management systems for storing and using classified information. For any one data system, however, most of these large expenses are either sunk costs or associated with specific individuals or billets.[20] People who have received security clearances can command a premium over equally trustworthy but uncleared individuals; paying such a premium is a cost to defense and intelligence budgets. The relationship between declassification and derating of billets is unknown; for instance, how much money would declassifying all information below the Top-Secret level save? All we know is that, if nothing were classified, there would be no need for vetting. But from an accounting perspective, that does not give us much to work with for assessing the costs of a particular classification decision. Nevertheless, if the marginal cost of classifying one more piece of data is low, even modest benefits would justify classification.

---

[20] ISOO, 2009b.

Focusing on financial costs presents a partial picture. Whenever information is restricted, that restriction affects how the information is used and may therefore reduce its value. Although opponents of classification tend to frame discussion in terms of broad information availability to support democracy, there are also legitimate arguments involving the value of information sharing to facilitate mission effectiveness and innovation. This applies particularly to GFM DI, whose very reason for existence is that the broad sharing and use of standardized force-structure information will yield significant cost savings and other benefits across DoD.

Keeping particular facts secret can reduce the operational effectiveness of other members of the military or the defense establishment.[21] It might be assumed that classifying information associated with GFM DI would not affect the effectiveness of U.S. military planners because everyone who needs to access the information should be cleared to work at the Secret level and should have access to a SIPRNET terminal. But "needs" is not the same as "will have the ability to," and the practical difficulties and introduced inefficiencies of being forced to use SIPRNET for analysis cannot be entirely ignored. Tangible examples of the difficulties include the recent rules put in place against the use of removable media in secure systems,[22] which have raised the difficulty of transferring even unclassified material from SIPRNET to NIPRNET or between two SIPRNET nodes without common file-sharing services. This also does not take into account the effects on many planning operations that might be done at the unclassified level or those requiring routine information sharing in unclassified environments, where some protection methods make it all but impossible to interoperate effectively with entities, such as some coalition partners, working at the unclassified level.

That an individual has a need to know a particular piece of information to do his or her job is, within established security procedures,

---

[21] This fits with the more general definition of costs as being a benefit lost.

[22] Jennifer H. Svan and David Allen, "DOD Bans the Use of Removable, Flash-Type Drives on All Government Computers," *Stars and Stripes* (Mideast ed.), November 21, 2008. These restrictions were subsequently modified.

part of demonstrating that they should be given access to a particular subset of classified information. Although need to know is a key to being provided access, having one does not guarantee that any specific individual will actually get access to that information. One consequence of certain types of information being kept Secret is that individuals, even those potentially authorized access to that information, may not know the information exists and may not, in fact, receive all the information they need to be most effective. Such a cost could be considerable.

All these impediments can be understood as minor *if* the planning processes that GFM DI supports were solely *hierarchical*—that is, if large problems were successively broken down into smaller ones, each of which were attacked one by one. A hierarchical process gives advance notice of which people need what information. People can thus be predesignated; being predesignated, they can be cleared, given SIPRNET accounts (if necessary), and granted access to the relevant information. Solutions from the lowest levels of the planning hierarchy are then brought to a higher level, integrated and deconflicted, and become the raw material for the next step. This process iterates until final integration is achieved to generate the plan. Hierarchical planning can work well in certain circumstances. Indeed, it is the primary method DoD uses to build its program objective memorandum (POM), the document that lays out military spending plans for the following five years in a piece-by-piece, step-by-step process. However—and this is a critical consideration—the POM results in a unified metric: a single topline number composed of many smaller numbers.

However, no organization as complex as DoD can do all its planning that way; therefore, it almost certainly will not be able to reliably predict who will need access to what kind of information or how often. Indeed, a large fraction of defense planning is necessarily *indicative*, which requires the broadest possible access to information. *Indicative* planning methods are most useful for horizontal planning involving the lateral coordination of elements throughout a complex organization.[23] In essence, the core of activity is a set of authoritative data that provides a touchstone for people to use in planning. The assumption behind indicative methods is that, if everyone takes sufficient account of the touchstone (which can be a future mark on the wall in addition to current status), each will make plans that happen to reinforce it.[24] Similarly, if planners start with the same information on what today's force structure really is, their plans will be more compatible than if they had had to make their own guesses about the force structure. Indicative planning methods are useful for planning the requirements of entities that are not in a position to tell one another what to do or to work closely with each counterpart to coordinate and deconflict their opinions about the future. Here, information, rather than command, is the coordination method. More broadly, indicative planning methods are more amenable to open-source-type solutions in which one is not certain of all the metrics.

In a hierarchical planning system, the hierarchy is central, and information is used to inform the process; personnel in hierarchies, as noted, can be centrally vetted for such information. With indicative planning methods, the information is central and the planners determine whether the information is important to them. In that sense, the planners are self-selected. As self-selected planners, they are not necessarily vetted, however. Indeed, it is typical that the overseers of any information set cannot predict who will benefit from its use. It is also not entirely obvious that every one of the planners will be within DoD. Congress is part of the planning process. Many planners who could

---

[23] Note that these are methods, not a system. This concept describes the French use of economic planning after World War II. The French government would put out GDP forecasts, which would give industrialists some guide to the size of the domestic market. This, in turn, would prompt them to invest accordingly in productive capacity, thereby helping France reach the GDP forecast. See Stephen Cohen, *Modern Capitalist Planning: The French Model*, London: Widenfeld and Nicolson, 1969.

[24] Analogously, if everyone believes, for instance, that Ethernet will be a core component of DoD's information architecture, they will tend to buy hardware and software that runs well with Ethernet, and, as a result, Ethernet will increasingly become the networking technology of choice.

benefit (and, in turn, benefit DoD) may work for the private sector, as contractors, third-party analysts, or suppliers.

Finally, plan testing and excursions—e.g., from the modeling and simulation community—are facilitated when the instruments they use can be run in the unclassified world.

Hierarchical planning systems are not incompatible per se with indicative planning methods, although they can be substitutes for one another to some extent. Hierarchical systems tend to be appropriate for internal parts of the planning domain, while indicative methods tend to be appropriate for the external parts. Even the POM process uses top-line budgetary guidelines and the Defense Planning Guidance—both of which are indicative elements—as well as a top-down process.

Conclusion

Justifying classification of a given piece of information requires a reason to believe that the damage from allowing the data to stand unclassified exceeds the damage from classifying it. In practice, measuring either kind of damage is difficult, so absolute assessments of the balance between them will likely not be possible.[25] However, doing more than simply worrying about how adversaries might hypothetically use particular types of data means that each side of this balance must be systematically examined and articulated clearly.

To reiterate briefly, we defined four criteria that must be met even before a classification argument can even be considered: (1) classification must reduce information flow to the adversary, (2) the data obtained must change what the adversary knows, (3) the knowledge must affect the adversary's decisions (albeit probabilistically), and (4) the decisions must damage the United States in some way. Even if it is hard to get more than just a hunch about which information affects which deci-

---

[25] Quist, 1993, Ch. 6, "Balancing Information-Disclosure Risks and Benefits," discusses issues in balancing costs and benefits in classification decisions, including the use of legal standards of evidence, such as "beyond a reasonable doubt," for thinking through what and how much certainty should be required to choose to classify (or declassify) a piece of national security information.

sions, the need to meet the criteria should force analysts to carefully think through what kinds of decisions adversaries might want to make, what information they may need from us to make the decisions, and the possible effects of the decisions on the United States.

Analysts cannot ignore the effects of classification on the performance of friendly forces. Better performance of DoD personnel is the central rationale for GFM DI—and such improvements could be considerable. Even if the financial costs of the classification system—the personnel vetting, information technology, and other investments that have already been made to allow the restriction and compartmentalization of information—are sunk costs, the efficiency losses are not. The burden from making such information more difficult to get and use and the barriers to innovation from restricting information of the sort contained within GFM DI are real if not precisely calculable.

CHAPTER THREE

The Global Force Management Data Initiative and Its Effects

This chapter describes GFM DI and how it changes the creation, storage, sharing, and use of force-management data.[1]

Force-Structure Information—Authorizations, On-Hand Forces, and Ready Forces

Force-structure information is a fundamental input for military decisionmaking. Military organizations must know the numbers and types of units, personnel, and equipment and how these forces are organized to make decisions. Designing force packages for specific missions today requires an understanding of where in the organization necessary forces are available. Planning forces for the future must be based on what is in place now and how future forces need to differ from current ones.

In thinking about force structures, three varieties of information are important. At the highest level are the number of forces *authorized* by law and how DoD implements it—since these determine the total size of the armed forces. Authorization numbers are usually larger than

---

[1] Discussion in this chapter is based on descriptive and policy information provided in GFM DI, *Concept of Operations*, Washington, D.C.: Joint Staff, Force Management (J-8), April 16, 2007a; GFM DI, *Organizational and Force Structure Construct* (OFSC), Washington, D.C.: Joint Staff, Force Management (J-8), October 12, 2007b; and informational briefings provided to the research team by Joint Staff Director for Force Structure, Resources, and Assessment (J-8).

on-hand forces, and should rarely be smaller, as monitored by property books and personnel systems. *On-hand forces* refers to the size and composition of personnel and equipment currently on the books—what Congress is willing to pay for. *Ready forces* refers to available on-hand forces, which can be no larger and are often smaller than on-hand forces. The difference between on-hand and ready forces is determined by the number of personnel on temporary duty or with temporary health- or injury-related disabilities and the amount of equipment in various states of repair. Authorized force-level numbers are therefore potentially twice removed from the force that is capable at any time of meeting the enemy. The current emphasis of GFM DI, and the focus of this analysis of security concerns, is on *authorized* force-structure information and *not* on-hand or ready forces (see Figure 3.1).

It is in the nature of authorization data to change slowly, although we were not provided precise statistics on how slowly. In general, most of the data valid in one year were valid the year before. Thus, our

Figure 3.1
Authorized, On-Hand and Ready Forces

What are you authorized?	What do you actually have?	What do you have to operate with?
Authorization data Authorized by law and organized by the components **Org servers**	"On-hand" data Property books and personnel systems	Readiness data Logistics and personnel systems

SOURCE: J-8/MASO, "Global Force Management Data Initiative (GFM DI): Vulnerability Assessment of Classified Data (VACDAT) Study," briefing, May 30, 2008.
RAND *MG989-3.1*

analysis could not rest on the proposition that purloined data would become obsolescent before it could be used.

The Problem GFM DI Is Intended to Solve

Because of DoD's complexity, management of force-structure information is quite a challenge. The military services manage their own force elements and subdivide them differently. The services used a variety of legacy systems for management. These were set up at different times for different purposes, evolved separately, and often used different data definitions and formats. As a result, the systems did not interoperate effectively, and there was no uniform, unambiguous way to identify people and things. In some cases, important information for decisionmaking—e.g., the command and support relationships between individual elements within the force—has not always been well documented or documented consistently and is therefore not transparent to decisionmakers in different parts of DoD.[2]

The fragmented nature of these legacy systems has meant that the effort required to understand the detailed makeup of the U.S. armed forces was difficult and time consuming. Because the legacy systems cannot exchange data easily, creating a consolidated force-management database would require a nontrivial amount of data translation. Keeping force-structure information current would require partially repeating that translation. The lack of uniformity and disaggregation meant that too much effort was required to answer fundamental questions about the potential to allocate forces for particular missions or deployments, which, if it did nothing else, complicated force planning. Figure 3.2 portrays the important uses of force-management data.

As a practical matter, it has also meant that different DoD functional organizations use different pictures of the force structure for

---

[2] See Joint Staff, *Capability Development Document for Global Force Management Data Initiative*, Washington, D.C., August 20, 2007, p. 10.

**Figure 3.2
Applications of Force-Structure
Information in Defense
Planning**

SOURCE: Adapted from J-8/MASO,
2008.
RAND *MG989-3.2*

planning, execution, and other activities, which introduces inconsis-
tencies and hurts management effectiveness.[3]

What the Initiative Is Doing to Solve It

GFM DI helps integrate service-authorized force-management data
by allowing users to access what used to be scattered heterogeneous
information as if it were one coherent database. It does this through a
series of mandates and standards that ultimately make disparate data
elements similar enough to be readily combined. In essence, therefore,
GFM DI has no data of its own; it is a portal to databases other organizations maintain (see Figure 3.3).

---

[3]   See discussion in GFM DI, 2007a, pp. 2–3, 5–6.

Figure 3.3
The Global Force Management Data Initiative System

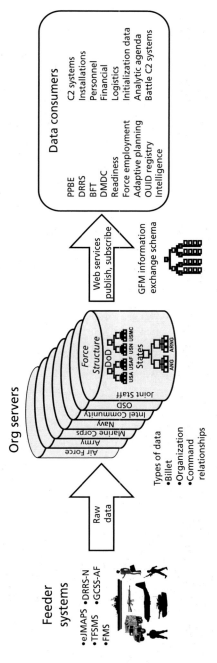

SOURCE: Adapted from J-8/MASO, 2008.
RAND *MG989-3.3*

For policy purposes—specifically, the issue of what should be classified—however, GFM DI is *not* a single database. If it were, its owners could mandate that certain fields, records, or attributes be classified.[4] Instead, because the initiative defines the rules for better interoperability with feeder databases and ways to combine force-structure information from existing systems, addressing any concerns requires (1) identifying security issues that result from either the broader sharing of data or the combination of different data types as a result of the initiative and then (2) addressing these problems by interacting with service data providers.[5]

What Changes Does GFM DI Require?

GFM DI is built on—allows users to draw from—extant systems, e.g., the Electronic Joint Manpower and Personnel System, Total Force-Structure Management System, Defense Readiness Reporting System–Navy, and Global Combat Support System–Air Force. The future architecture (Figure 3.3) assumes real or virtual consolidation of the many servers that hold force-management data to a set of what are called *organization (org) servers*. The expectation is that org servers will be consolidated at the service level (thus yielding four total), with one additional org server each for the Intelligence Community,

---

[4] A *record* is a set of data about a particular person, place, or thing. A *field* is a characteristic of a person, place, or thing. An *attribute* is a value that a field can take. A *table* is an ordered collection of records. A *database* is a collection of linked tables. Thus, everything in a table about Sergeant Snuffy Smith is a record. "Security clearance" is a field. The fact that a given individual has a Top-Secret clearance is an attribute. One could conceivably declare that information about Sergeant Smith is withheld from the database (a record). Or one could declare that information about a person's security clearance not be in the database (a field). Or could declare that all clearances higher than Secret should be shown as just Secret (an attribute).

[5] To the extent that security concerns are identified after they arise—the problem is discovered only after a service has provided the data and GFM DI–compliant systems have shared it—they can only be addressed after the fact, by removing the relevant information. If the data leaks to adversary hackers before it is removed, the act of removal should affect only future data updates, rather than the data itself.

the Joint Staff, and the Office of the Secretary of Defense (OSD). No one org server contains all the information that every user might want. Requests that one org server cannot satisfy are forwarded to the one that has it; when the data come back, the original org server bundles the entire answer in a package and presents it to the user. All this, perhaps needless to add, requires a great deal of interoperability. The changes included in GFM DI are as follows:

- GFM DI establishes a data dictionary, one based on NATO's Joint Consultation, Command, and Control Information Exchange Data Model.[6] This standardizes the meaning of a particular field and establishes the legal values that attributes in the field can take. Although the parent data dictionary includes a wide variety of options, the business rules of GFM DI substantially reduce the level of detail required for compliance with the initiative. For example, a unit containing an aircraft only has to list the aircraft type, although the dictionary contains many additional descriptors for aircraft. Furthermore, although data dictionaries tend toward explicitness, this one allows, indeed sometimes encourages or even requires, blurring certain types of data. For example, in some cases, data providers are directed to hide potentially sensitive attributes under the general label of "not otherwise specified (NOS)" or other general nonsensitive category codes. The presence of such generic codes make it possible to satisfy the requirements of the initiative while still concealing potentially sensitive information about a platform.[7]
- GFM DI specifies a set of minimum data elements—a common set of fields that should be filled in for every record of a given type. For example, all billet data must include rank and grade, occupational code, component (e.g., Air Force), skill level, service, effective date, and service and billet identification (optional beyond GFM DI organization unique identifier). But the entry

---

[6] See the Multilateral Interoperability Programme website for further details.

[7] See the appendix for examples of aircraft types and additional descriptions.

need not contain any other information. Other force-structure elements have their own minimum information requirements.

- Although the minimum set of essential data for each force-structure element is quite limited, the initiative does require that force-structure entries be complete. Because of the hierarchy inherent in command and support relationships and the requirement force-structure elements add up to authorized levels, any discretion that military services might have had to leave records out of some systems that are within the mandatory categories *for entities that are not in of themselves classified* would no longer exist.

- It mandates the common-access protocols, which permit a request broker (software that converts a user request into software code) to work the same regardless of which org server it is communicating with. This also facilitates server-to-server data exchange.

- It mandates the expression of data in Extensible Markup Language (XML) and provides a uniform schema that governs how XML will be used. XML is a standard web syntax that help can overcome seemingly small but critical differences in the way database engineers from different vendors organize their data.

- It establishes a uniform identifier, the force module identifier (FMID), for every billet, place, or thing that can index a legitimate record in the minimum data set (technically, an FMID is for every element of the GFM XML Schema Definition). In essence, everyone and everything gets a number. This number, in turn, can (and with time may well) become the standard reference for other information collections. Thus, a personnel database (external to GFM DI) may report that Sergeant Snuffy Smith currently fills billet 928-37928-47923, which, in turn, can be found in GFM DI; this database then provides information about what is authorized for such a billet.

- GFM DI establishes a hierarchical framework to collect and organize force-structure information for joint integration. To support this, it provides the rules to create high-resolution force-structure trees that span the "forest" from the topmost echelon, across the unit identification code boundary, down to individual billets. This provides uniform ways to represent command and support relationships among billets, units, buildings, and other entities. In the current phase of GFM DI, only administrative and default relationships are planned for the unclassified implementation of GFM DI.[8]

What Additional Changes Might GFM DI Implementation Produce?

The changes above describe the data format, content, and sharing GFM DI standards require. Although such changes were central to our examination, we identified two other potential changes that may occur in response GFM DI, notably in force-structure information and in the databases that feed GFM DI.

Potential Changes in Force-Structure Information

GFM DI mandates maintaining and sharing a minimum set of essential data for each element of the U.S. force structure but does not set a maximum. Each service (or contributing component) can establish tables and fields of its own choosing within its own structure, potentially greatly increasing the information on force structure included in GFM DI–compliant systems beyond the minimum data set (and potentially beyond such information contained in the legacy feeder systems). Some of the services already have. Such fields may or may not be the same as those other services use. Users may be given access to such records, but if the tables and fields are not in the data dictionary—and there is no rule that they must be—they cannot be shared and combined via GFM DI–associated data-sharing systems.

One primary way to add data to GFM DI systems is to use *alias fields*. Alias fields were added to GFM DI data structure to allow users to include other names for force-structure elements (e.g., an additional numerical designation for a platform, an additional title for a billet). Alias fields can also, however, be used to add descriptive elements to a force-structure element that the minimum data set does not accommo-

---

[8]   Joint Staff, 2007, p. 23.

date (as long as they conform to an agreed set of alias types, such as the Global Status of Resources and Training System Long Name). These fields provide flexibility and allow customization of the system to meet the needs of different DoD users without centralized action. However, this flexibility also raises the potential for information to be added that is sensitive if not already classified.

Although GFM DI does not place any controls on what other information is included in org servers (so sensitive information could be added to them beyond the minimum data set required by the standards), so do many other information systems, and the latter can disclose the same data. So, although we examined the implications of potential disclosure of U.S. force-structure information beyond the minimum data set in our subsequent analysis, implementation of GFM DI disclosures constitutes *potential* rather than *definite* security concerns.

Possible Changes to Databases

GFM DI is not itself a database and does not require creation of new databases of force-structure information. In theory, GFM DI's requirements can be met without changing any database, except insofar as necessary to insert fields and attributes necessary to meet the far-from-onerous requirements of the minimum data set. Otherwise, it would suffice to put a translator and broker between each database and GFM DI to render invisible any differences between the data requirements of GFM DI and the content of the source databases. The mandates of GFM DI apply only to what users using GFM DI–compliant brokers see, not the contents of the org servers themselves.

In practice, GFM DI implementation could force changes in such databases. It is wasteful for an organization to maintain two databases doing similar things, and writing a translator consistent with the original database but compliant with GFM DI mandates creates just one more place for error to lurk. So, the services have been transforming their databases to be more natively compliant with GFM DI. At the very least, many services have consolidated their previously disparate databases into one or a few servicewide databases that are intended to give the services themselves a better picture of their own force structures. A notable example is the U.S. Navy's effort to combine many different data repositories (databases, spreadsheets, flat files) into one servicewide database.

Because GFM DI will not centralize force-structure information, it is meaningless to talk about hackers gaining access "to GFM DI" and therefore obtaining the entirety of U.S. force-structure information. Hackers can, however, get into org servers. They could also pose as legitimate users and thereby download arbitrary portions of data accessible from GFM DI. The greater interoperability GFM DI will create among existing databases, the building of publishing and messaging systems for sharing force-structure information, and the broader use of uniform force-structure information across DoD are likely to mean that more (and "better") force-structure information will be stored in more places. As a result, these changes could increase the availability of this sort of information to adversaries with the ability to penetrate DoD networks.

CHAPTER FOUR

Potential Security Concerns and the Possible Benefits of Classification

Do the data and its aggregation spurred by GFM DI pose security concerns? To address this question, we identified security concerns that had been raised in connection with GFM DI and systematically applied the four criteria of Chapter Two to determine whether there was any basis for recommending that all or part of GFM DI data be classified or otherwise restricted (including being made less specific).

These security concerns centered on the potential for adversaries of different types—states and nonstate actors—to use GFM DI data. We distilled the concerns into the following three questions:

- Will GFM DI provide adversaries information about U.S. force structure they could not otherwise obtain?
- Will GFM DI make it easier for adversaries to confirm information they think they already have about U.S. forces?
- Will the potential for GFM DI to make it easier for adversaries to aggregate our force-structure information create new security concerns?

The first two questions arise from the second criterion: The first asks what the adversaries know (as a result of access to GFM DI), and the other asks how confidently they know what they know (again, as a result of access to GFM DI). The last question reflects the second criterion (what they would know as a result of data aggregation) and the third criterion (how the knowledge would change what they do and thereby raise security concerns).

By way of introduction, note that a classified force structure currently exists and will continue to be classified under GFM DI. The issue here is classifying currently unclassified force-structure information that might be sensitive due to its aggregation.

Data aggregation presents a two-part concern. One is the possibility that the standardization GFM DI creates will improve the ability of adversaries to weave a "big picture" view of U.S. force structure (a "mass aggregation" concern). The second is the possibility that the transparent linkage of individual elements within the force structure (e.g., billets to units or platforms through command or other relationships) will reveal new information about one or the other (a "microaggregation" concern).

For each security concern, we examined how the changes required or possibly spurred by GFM DI would affect potential adversaries' access to this data. We then examined the data and asked, for each concern, whether the four criteria laid out in Chapter Two supported the classification or other restrictions on the data. We examined adversaries in two broad classes (state and nonstate actors) and examined three general classes of decisions relevant to force-management data (force employment, force planning, and intelligence decisions). The following sections address each security concern in turn.

The fundamental basis for this examination was the minimum data set that GFM DI mandates be maintained and shared. Because of the flexibility associated with the initiative and the military services' implementation of it (see Chapter Three), the service providers are also free to supply additional force-structure information that would similarly be shared. This flexibility makes it impossible to make a wholly comprehensive assessment. Without knowing how data providers will customize their databases or what information they might include in alias fields, no definitive statement is possible. This is why we first made more-definitive judgments about the data included in the minimum data set GFM DI requires, then laid out the *potential* security concerns that might arise from GFM DI.

Will GFM DI Provide Adversaries New Information They Would Not Otherwise Have?

The broadest security concern about GFM DI is that it could provide adversaries hitherto unknown information about U.S. force structure: both an overall picture of U.S. forces and information about each of its elements.

This concern and whether it alone justifies classifying or restricting data touches on our first two criteria: (1) that classifying the data (i.e., moving it from NIPRNET to SIPRNET) would reduce adversary access to such data and (2) that they have not acquired such data from other sources. Judging by these two criteria, our assessment is that, in general, the information included in the GFM DI minimum data set *will not provide adversaries with new information that they could not get elsewhere.* The design of the data system means that such situations could, however, arise in implementation and merit some ongoing attention. In the following subsections, we first examine whether classification would reduce adversary access to information, then turn to how access to this information would affect their knowledge.

Would Classifying Some or All of GFM DI Actually Reduce an Adversary's Access to Information?

Whether classifying some or all of GFM DI would actually reduce actual or potential adversary access to the data is sensitive to the specific type of adversary. Currently, the data systems under GFM DI are already largely on NIPRNET, as unclassified GFM DI will be. Although NIPRNET is unclassified, posting information on it is not the same as making that information broadly available on the open, public Internet. Various access control measures protect NIPRNET by limiting access to certain areas. A decision to classify some or all the GFM DI data would move that information from NIPRNET to SIPRNET, a network with additional protections and access controls that provides a higher level of security for classified data.

In asking whether that change would deny an adversary anything, the first question is whether that adversary is currently capable of penetrating NIPRNET. It has been publicly reported that NIPRNET, even with its protective measures, has indeed been penetrated.[1] Thus far, the penetrations have been linked to competitor states. In fact, it is probably safe to assume that the intelligence agencies of many large countries have attempted to penetrate NIPRNET and have enjoyed some measure of success.[2] Moving data from NIPRNET to SIPRNET could potentially reduce the amount of that data that adversaries with that capability would be able to collect. However, not all adversaries are in a position to hack into intermediate-security computer networks like NIPRNET. Many nonstate groups, either because of a lack of interest in or lack of success with such capabilities, will not be able to penetrate NIPRNET at all. For them, it is important to understand that the shift from NIPRNET to SIPRNET will have no security benefit, not because it would not secure the data "better" but because the added security is currently unnecessary. For these possible attackers, the choice to classify any of these data does not meet our first criterion.

However, the assumption that some states have already penetrated NIPRNET also has implications for the decision about whether to move this information to SIPRNET. The systems feeding information to GFM DI that are already on NIPRNET may only require data reformatting and reposting to render them GFM DI compliant. The information itself will still be on NIPRNET, and its availability will not have changed. Adversary states interested in this sort of information that have already exploited the network have also likely already accessed the information in question. Although force-structure information does change over time, as strategies, resources, and other variables shift, much of the data remains quite constant.

The important consideration, then, is not denial of any force-structure information, but the difference between total current data via GFM DI and what adversaries may have derived from individual component databases. Without actually knowing what any adversary

---

[1]   Robert Marquand and Ben Arnoldy, "China Emerges as Leader in Cyberwarfare," *Christian Science Monitor*, September 14, 2007, p. 1; Vernon Loeb, "NSA Adviser Says Cyber-Assaults on Pentagon Persist with Few Clues," *Washington Post*, May 7, 2001, p. A02.

[2]   This would include freelancers either operating under the direction of a government or hoping to sell stolen data to a government.

has already collected, it is impossible to know for sure how much we would be denying them by moving the GFM DI systems to SIPRNET. It is likely foolish to blithely assume that they have none of the data relevant to GFM DI, but it would be unduly pessimistic to be sure that they have it all. A great deal depends on the percentage of information available on NIPRNET that any one country has gathered; the priority the country would accord to GFM DI data with respect to other, similar data it has harvested from other sites;[3] and other, hard-to-predict circumstances (such as whether GFM DI hosts are easy or hard to crack). It is not impossible that countries have copies of the data but do not realize it (or realize it, but do not understand its utility) if they have stolen sufficient data but lack the resources to analyze and utilize it all.[4] As a result, for state adversaries, the choice to classify may meet our first criterion but only prospectively.

The odds that smaller countries have such data are probably lower than the odds that Russia or China do, although the smaller countries may benefit from sharing intelligence with the latter. Penetrating the NIPRNET is not trivial. In the cyberintelligence business, the size of the potential attacking force matters. It takes large numbers of "bodies" to carry out multiple penetrations and insert multiple implants. The larger the population of educated people in the country, the greater the number of exceptionally high-grade hackers and the wider pool of talent from which a state intelligence agency can draw. The odds that a country has access to force-authorization data will be roughly proportional to its effective size (measured in terms of university students).

---

[3] Web spiders (computer programs that browse the web automatically and methodically) are reasonably good at gathering a large amount of data in HTML, but are less appropriate for gathering similar amounts of database material, particularly material that can only be accessed via queries. Technologies in development may greatly improve the ability of spiders to use other types of data repositories.

[4] This is a common problem in intelligence organizations, particularly those with access to broad electronic collection resources. See, for example, discussion of the National Security Agency's challenges in metabolizing the volume of communications it intercepts daily (for example, discussion in Matthew M. Aid, "All Glory Is Fleeting: SIGINT and the Fight Against Terrorism," *Intelligence and National Security*, Vol. 18, No. 4, 2003, pp. 72–120).

The odds that a terrorist group would have access to GFM DI are even below those of the smaller nation-states. Such groups are small, and while some members are technically sophisticated (e.g., Irhabi007, who established and maintained jihadist propaganda servers) and talk about computer hacking, the evidence that they are capable of serious penetration is pretty thin. There has been no *reported* case of a terrorist group getting into a system that is as well protected as NIPRNET is. One might imagine that Hezbollah, which is very close to Iran, might have U.S. force-authorization data that Iranian hackers had picked up (if any) that was deemed releasable and relevant, but even that is not certain. Most terrorist groups (notably al Qaeda) lack such close relationships with capable state supporters.

This leads to a broad conclusion: As long as GFM DI data are hosted exclusively on NIPRNET, the risks of keeping GFM DI unclassified are roughly proportional to the capabilities of any organizations that can make good use of the data. In particular, terrorist groups without state sponsorship are unlikely to be able to acquire such information through computer hacking.[5] Thus, the case for classifying data elements that are of particular interest only to terrorist groups is weakened to the extent that terrorist groups cannot not pull the information from NIPRNET.

Another implication of GFM DI data's residence on NIPRNET, rather than on the web, is that access to it is likely to be episodic, perhaps even unique, rather than continuous. True, hackers can and do put implants on computers that allow them to receive and transfer periodic updates of the computer's contents. Thus, it is not impossible that someone can get near real-time feeds either from one of the various contributing servers for GFM DI or from users who access GFM DI data often enough. But without such devices (whose existence is unknown), stolen GFM DI data are apt to be a bit dated. In particu-

---

[5]   When human intelligence, rather than computer hacking, is the approach vector, size may be less important and contacts and religious or ideological sympathy more important. Thus, terrorist groups may be more competitive than states. To the extent that most NIPRNET users have access to SIPRNET, data are scarcely any safer from insiders on SIPRNET than NIPRNET, except that very large data transfers from SIPRNET can be more difficult to carry out.

lar, the quality of information available to authorized users is likely to exceed that available to thieves. Does this matter? The best answer is, probably, "a little." Authorization data are relatively static; they represent plans and programs, in contrast to actual assets ready to go. Over a ten-year period, it is more likely that the person filling a billet will change than that the billet itself will change in key respects. However, since part of the sensitivity of GFM DI information is due to its potential links to other databases (e.g., personnel records), as explained later in the chapter, the dynamic nature of the latter have to be taken into account to understand the risks from choosing not to classify GFM DI data.

Finally, the distinction between NIPRNET and the web is related to the effect that GFM DI process has had on data accessibility, notably the consolidation of service databases. These consolidations, in turn, have likely reduced the work hackers take to find the systems housing the data. Individual sites, for instance, are probably easier to find because there are more references to a consolidated site than to each of its components. Getting a complete picture now requires compromising fewer sites, although each of the major sites might be much better protected than the minor sites were.

In sum, it would be an exaggeration to claim that classifying GFM DI information would make all the difference in the world to what adversaries can get. Low-tech adversaries (e.g., terrorists, developing states) may not be able to penetrate NIPRNET as it is. High-tech adversaries may be kept from the information, unless they have other ways of penetrating SIPRNET and view the information as sufficiently important to make the effort to get it.

Assuming an Adversary Gains Access to GFM DI, Would It Provide Force-Structure Information It Could Not Get Elsewhere?

Setting aside questions about what type of adversary could gain access to what classes of computer networks, assume that adversaries will get the data. Will it tell them anything they do not already know or could not find easily elsewhere? Exploring this issue requires looking at the data fields and assessing alternative ways relevant adversaries could get that information.[6] This discussion focuses on our second criterion: whether the information will affect an adversary's level of knowledge.

Compliance with GFM DI does not require disclosure of particularly sensitive information. In the minimum data set, as laid out in the GFM DI data dictionary, many fields are optional, and the codings used are restricted to a subset of possible codes. Data creators are often directed to use a single generic option (e.g., "aircraft, not otherwise specified") that provides little additional insight into any one unit, platform, or organization, other than that it exists.

Most of the information specified in the minimum data set— the identity and enumeration of billets, platforms, units, and organizations—can also be acquired by other means. Analysts have known for decades that published and other open-literature sources can inform those who would count and characterize others' military organizations. Indeed, doctrinal and technical publications by U.S. practitioners of open-source intelligence (OSINT) openly discuss what types of information are available from such sources.[7] Open sources have been used to map organizational structures, force structures, and other contents of tables of organization and equipment as far back as the Cold War (not to mention from warfare's inception) and have been a key element of U.S. intelligence activities. The challenge to build a comprehensive picture of military forces, organization, and orders of battle mainly entails weaving together information from many open sources, cross validating it, and building the unified picture needed for opposing-

---

[6] This essentially asks the "worst case" question about the content of GFM DI: Starting from the premise that all protective attempts are ineffective, what is the downside of full compromise of the information? This is an upper-end estimate of the potential damage adversaries might do with the data. and therefore an upper-end estimate of the benefits of classifying it. In the real world, where protective measures will undoubtedly be effective in some cases against some adversaries, the actual benefits will be less.

[7] See, for example, the October–December 2005 issue of *Military Intelligence*, published by the U.S. Army Intelligence Center and Fort Huachuca, which examines OSINT as a part of military intelligence from a number of directions. In particular, see Donald L. Madill, "Producing Intelligence from Open Sources," *Military Intelligence Professional Bulletin*, Vol. 31, No. 4, October–December 2005, pp. 19–26, which discusses the role and utility of OSINT for these types of information-gathering requirements and its relationship to classified intelligence.

force analyses. Such a challenge is readily met. Consider, for example, this quote:

> [In 1979,] TRADOC tasked DIA to produce a definitive unclassified intelligence document on the Soviet Army. This multivolume document was to serve as a comprehensive baseline reference to support threat instruction and training at TRADOC schools and centers, as well as the Opposing Force (OPFOR) Program throughout the Army. . . . This collaborative production effort culminated in 1982, with the published coordinating draft of FM 100-2 in three volumes. . . . Many customers, as well as analysts in the Army's intelligence production centers, were surprised by the manuals' thoroughness and accuracy. In many cases, they learned that essentially the same information they were accustomed to seeing only at the sensitive compartmented information (SCI) or collateral classified levels was also available from open sources. Moreover, these unclassified products could enjoy wide dissemination throughout the Army and other services. (Madill, 2005, p. 20)

These U.S. Army field manuals on the Soviet Army circa 1984 show content strikingly similar to that covered by GFM DI.[8] They document important details of Soviet force structure, nominal command relationships, manpower, and general force characteristics of a military from a country that was far more secretive and closed than the United States and during a time when there were many fewer open sources (commercial sources, the Internet, etc.).[9] More-recent examples abound; RAND's examination of China's People's Liberation Army

---

[8] The version we reviewed was U.S. Department of the Army, "The Soviet Army: Troops, Organization, and Equipment," FM 100-2-3, June 1991, listed as superseding the version published July 16, 1984.

[9] See Stephen C. Mercado, "Sailing the Sea of OSINT in the Information Age," *Studies in Intelligence*, Vol. 48, No. 3, 2004, pp. 45–55. The author was an analyst in the CIA Directorate of Science and Technology. The article includes a telling quote: "The world today abounds in open information to an extent unimaginable to intelligence officers of the Cold War," suggesting a similar construction of force-structure information today for professional, well-resourced intelligence analysts would be straightforward.

was built from open or gray sources, and China devotes great energy to restricting information.[10]

Since then, the richness and the detail of the information on militaries available from open sources have only grown; such sources have become far easier to get and use. Commercial services, such as Jane's and Periscope, cover military forces and platforms with fine resolution.[11] These services provide information on crew complements, unit associations, and other information in comparable or better detail than what GFM DI offers. While the data initiative covers only authorized force structure, commercial services seek to provide insight into at least on-hand forces, if not estimates on force readiness. Individual websites—including such communal "crowd sourced" information repositories as Wikipedia—aggregate open-source information into force structure and order-of-battle mappings. Documents that are sources for information in GFM DI (e.g., the U.S. Army Table of Organization and Equipment) have versions available on the open Internet.[12] Databases even exist that provide specific data on individual U.S. military planes.[13] Other sites provide order-of-battle information for deployed forces (e.g., for current operations in Iraq)[14] and comprehensive military balance publications include broad information on force deployment numbers, although not in real time.[15] Even more detailed data

---

[10] James C. Mulvenon and Andrew N. D. Yang, eds., "The People's Liberation Army as Organization: Reference Volume, v1.0," Santa Monica, Calif.: RAND Corporation, CF-182, 2002.

[11] Jane's is a British publishing house and website that specializes in military and intelligence issues. Military Periscope.com is an online database initially developed by the U.S. Naval Institute.

[12] U.S. Army, "Table of Organization and Equipment," March 8, 2000.

[13] Dutch Aviation Society, "United States Air Force & Army (1948–Now)," *Scramble*, undated.

[14] John Pike, "US Forces Order of Battle," GlobalSecurity.org, May 19, 2008.

[15] We consulted James Hackett, Nigel Adderley, Andrew Brookes, Jason Alderwick, Mark Stoker, and Hanna Ucko, "The Military Balance 2007," International Institute for Strategic Studies, 2007. Others provide information on naval positions with more-regular updates (e.g., STRATFOR Global Intelligence, naval update map, web page, April 1, 2009).

can be gathered from perishable open sources.[16] Still others have made credible efforts to catalog and describe even classified programs.[17]

Indeed, recent U.S. military doctrine reflects the value of such open sources for answering these types of questions about other nations' military forces. A recent intelligence field manual, now on the web, specifically directs that collectors

> conduct periodic searches of webpages and databases for content on *military order of battle (OB), personalities, and equipment.* Collecting webpage content and links can provide useful information about *relationships between individuals and organizations.* Properly focused, collecting and processing publicly available information from Internet sites can support understanding of the operational environment.[18]

An assessment of Internet sources for the use of U.S. military professionals similarly assesses them as good sources for "force structure

---

[16] For example, when assessing concerns about the embedding of media members with military units, one observer stated that

> Furthermore, in an age where much military information can be gleaned through open sources, secrecy has become increasingly difficult to sustain. Commercial satellite imagery, cellular and satellite telephone intercepts, and the Internet can all be employed to track the movements of military forces. A fusion of that information, combined with the full-page war maps of *The New York Times* and the retired generals' analysis on television, could have provided the Iraqi military an accurate picture of the US dispositions. (Brendan R. McLane, "Reporting from the Sandstorm: An Appraisal of Embedding," *Parameters*, Spring 2004, pp. 77–88)

[17] Jonathan Pike, then at the Federal of American Scientists, had developed a fairly detailed listing of U.S. space capabilities, again, from unclassified sources. See also William Arkin, *Code Names*, Hanover, N.H.: Steerforth Press, 2005, for lists of intelligence programs (with varying degrees of accuracy) complied, similarly, from open sources.

[18] U.S. Department of the Army, "Open Source Intelligence," FMI 2-22.9, December 2006 (expired December 2008), pp. 2-5, 2-6 (emphasis added). Reflecting the open availability of information in the current era, this document is available on the open Internet, even though it was marked "For Official Use Only."

tables and budget figures, from military maps and photos to weapons and equipment capabilities."[19]

Potential adversaries can gather information from other places. Overhead imagery makes it possible to count some platforms and track their locations. Information the military publishes itself (e.g., on unit websites), catalogs of organizational structures, and even direct observation of individuals—through techniques as mundane as reading their uniform insignia at public events—can provide insight into their activities and, by association, the programs and units of which they are a part. As with OSINT in general, using such sources requires weaving together disparate and fragmentary information into a coherent whole. Clearly, aggregation and integration efforts are well within the capabilities of any state adversary of concern.[20] Thus, arguments to classify part or all of GFM DI does not meet our second criterion: affecting what an adversary knows.[21]

That most of the information GFM DI covers is readily available through other sources, does not mean that *some* sensitive information is not being included in GFM DI. GFM DI's minimum data set and data dictionary limit the fields and attributes in the database, but users can add additional information using the alias fields. These fields are meant for alternative names for force-structure elements, but they can be used for additional elements. The only restriction is that the alias fields must refer to indexed entities in the various data tables. Thus, using alias fields, a data contributor could enter classified or contextually sensitive information into GFM DI. Adding the security clearance associated with a billet would be an example; this information is sensitive both because it is potentially useful to an adversary and because it is not readily observable through other means. Someone reading these

---

[19] Ed Metz, "Capturing Military Information on the Web and Elsewhere," *Online*, September 1, 2004, pp. 35–39. The author is a reference librarian at the Combined Arms Research Library of the U.S. Army Command and General Staff College.

[20] Although nonstate actors have also used such intelligence gathering, this kind of "big picture" force structure elucidation is largely irrelevant to most of their activities.

[21] This portion of our analysis led to the generalization: the more readily observable data is by other means and other sources, and the lower the risk to the adversary of using those other sources to get the same information, in general the weaker the case for classifying the data.

data could infer which billets were more sensitive than others, thereby helping to identify which individuals to try to recruit or exploit.

As a rule, detailed information on individual billets (e.g., skill or clearance level) is less available in the open sources and harder to observe directly. Yet it can also sometimes be inferred from open sources. For example, Trevor Paglen gathered copious information on classified aeronautics development and testing units using "resume intelligence"—gathering information on units by examining the resumes of individuals to trace their careers backward to a likely association with a particular unit.[22] A similar approach could unearth similar information about personnel in units that are not especially sensitive. In contrast to less-detailed force-structure information, for which a complete picture may exist in open sources, data on individual billets or specific units in open sources is almost always fragmentary. As a result, use of alias fields could add information to GFM DI–compliant systems that adversaries could not easily get through other means.

The possibility of leaking sensitive data is not something unique to GFM DI. Any open-ended database and almost all websites and emails may allow unsupervised users to disclose sensitive data. Also, because data contributors are still figuring out how they will comply with GFM DI, this is a potential rather than a proven problem. In some cases— e.g., required clearance levels—the information would pass the four criteria: Classification could hide the data from those interested in it; it might add to what adversaries know; the knowledge could inform decisions; and the decisions could damage the United States. Still, there is a wider variety of information that could be entered into such fields that would fail one or more criteria but be valuable for system users to include. These distinctions are important for considering prudent responses to this potential security risk, which we will take up in the next chapter.

---

[22] Trevor Paglen, *Blank Spots on the Map: The Dark Geography of the Pentagon's Secret World*, New York: E.P. Dutton Adult, 2009.

Will GFM DI Make It Easier for Adversaries to Confirm Information They Think They Already Know?

If adversaries are not sure that what they know is correct, access to an authoritative data set against which to validate or invalidate their data may give them information that they did not have before (and, conversely, restricting access to it would deny them this confidence). It is less certain whether that additional information actually affects their level of *knowledge,* rather than just their stock of *data.*

Even though most adversaries already have a basic understanding of U.S. force structure, security proponents may argue that GFM DI information might still help them because of this opportunity to validate what they know against a U.S. government source. The fact that the latter is used in U.S. decisionmaking would supposedly increase its value in the eyes of adversary analysts and decisionmakers. This argument would apply more to planned or projected force-structure information (i.e., authorizations projected several years in the future) than to current force-structure values. If this is indeed the case, this argument would meet our second criterion for classification—that access to the data would augment their knowledge.

But whether GFM DI will be viewed as a special or particularly valuable data source—it *is* U.S. government information and, hence, more trustworthy or useful—may vary from adversary to adversary. The idea assumes they would view these data as U.S. DoD users would, rather than as a deceptive plant meant to fool unauthorized users. Adversaries might instead institutionally view data they themselves had developed from a variety of sources as more valuable and accurate—either for valid reasons (e.g., it *is* "better" because it is more up to date than a GFM DI–compliant system hacked to acquire additional data) or idiosyncratic ones (e.g., the organization has a large investment in the activities producing their own force-structure information and not using it would not serve their internal organizational interests).

The argument that the authoritative nature of GFM DI data provides useful validation and thus justifies classification may in fact *fail* our second criterion, or it may fail one of the last two criteria—whether the change in knowledge would change decisions and whether those decisions would necessarily run against our interests. Here, it matters what adversaries know and what they plan to do with what they know. The better the picture a nation or nonstate group has initially, the less value they get from validation. Furthermore, precise force-structure information relates in different ways to the decisions of different potential adversaries. To illustrate these differences, we will discuss state and nonstate adversaries separately.

Adversary or Competitor States

Most states of significant military concern have a reasonably good picture of U.S. force structure even in the absence of any information that would be included in GFM DI–compliant systems in the future. As a result, these systems would, at most, be secondary sources of information that might help an adversary refine their understanding. *How much* refinement would depend on how uncertain the adversary's initial estimates are. If the adversary comes to GFM DI with a great uncertainty about a factor ("I think they have 20 tank battalions, but it could be anywhere from 10 to 30"), access to GFM DI could reduce that uncertainty considerably. But most state adversaries are unlikely to be that uncertain when it comes to U.S. force structure. Their estimates are likely to be mostly correct, even after expending relatively little effort, so the value of subsequent error correction is likely to be modest. To make this point mathematically, if you have one estimate with a root-mean-square (RMS) error term of 1 and introduce an independent estimate with an RMS error term of 2, both readings taken together are tantamount to an estimate with an error term of roughly 0.9 (and only if the readings are, in fact, independent of one another). This is an improvement but not a great one. As a result, in the majority of cases in which GFM DI–covered data will improve the precision of an adversary's knowledge of overall U.S. force structure only modestly, the result will only barely meet our second criterion.

But whether or not an improvement in an adversary's knowledge is modest in terms of error reduction, our third and fourth criteria still apply, and we have to ask whether that modest increase in precision is important. This requires asking whether there are decisions that can be changed in the presence of slightly more-precise data (and whether such decisions are adverse to our security).

The first point to make is that, for many adversary decisions of potential concern, the data GFM DI covers—authorization information—are simply not that useful. For example, GFM DI information will generally *not* be helpful to adversary warfighters in the field. It does not help much with target selection or operational decisions because the information GFM DI requires does not disclose the *significance* of one piece of information relative to others. It does not help with targeting itself because authorized data do not say very much about where anything is likely to be at the time of attack.[23]

As a result, because of the nature of GFM DI, it is most relevant to consider how adversaries might use the information for planning. The value of GFM DI data to their planning efforts, if any, would be related to the inferences they would make about U.S. capabilities and even strategy from looking at U.S. force organization and composition. This can only be a guess on their side, since a great deal of what they "learn" is a function of what they did not know before they received the information. Furthermore, what they actually learned would depend on the difference between what they conclude about U.S. force structure from the GFM DI and what they would have surmised about U.S. force structured based on what they know about U.S. military assets—something to which military analysts worldwide are privy.

---

[23] In the field (and even at home), units are frequently organized into task forces that commanders construct from different force elements (e.g., an extra tank company, special section attached, aircraft) to meet particular missions. Although task forces may consist of force elements associated with their primary units, the differences in the units' mixes may be very important, e.g., an extra brigade or division-level element, a specialized air element (brigades and wings function best when deployed cohesively, but there are enough mixing and matching exercises that any planner needs to be concerned about this issue). Confusion and deception could add further variety. Confusion often emerges from incorrect associating of the unit by virtue of a detachment member simply being present. Deception may arise if individuals intentionally present misleading associations (e.g., having the wrong unit patch and insignia on a uniform). Other, more-prosaic differences merit note. Many units are not at full strength. Sometimes equipment has not been delivered, is down for maintenance, or cannot function. People may be missing because of personnel pipeline issues, temporary duty elsewhere, or detachments being stripped off. Any hostile planner should be concerned about these differences.

The standardization in the U.S. military—and militaries in general—means the a priori expectations of observers are likely to be pretty accurate. Most U.S. units are templated to a high degree. This means that a brigade, wing, or other standard unit will typically consist of a particular mix of elements (vehicles and personnel) and that units of that same type will share most of the same characteristics. While GFM DI provides a detailed template of the unit in question, a good approximation would have been available from other sources in terms of authorized data. There are only so many logical ways of constructing a tank battalion, for example, and the differences among battalions are likely to reflect some obvious features (e.g., the exact kind of tanks, how are they supported) before they reflect the more subtle ones. There is also the danger to foreign analysts of the U.S. military of reading too much into such differences. They may not represent hidden doctrine as much as exigency, contingency, bureaucratic politics, personalities, microideologies, variations in budgets, and other constraints.

It is hard to believe that analyzing GFM DI for information about U.S. doctrine or future plans would provide anyone with any significant value. Americans carry out their debates in public, leaving the rationales for much of what can be observed in fairly plain view. Every service college has its journal, as do many warfare communities and even intelligence agencies (e.g., the National Geospatial-Intelligence Agency). Defense manufacturers tout their ideas in print, and there is a lively subscription-based press devoted to defense and aerospace matters. The daily press, television, and radio all have articles on military affairs. Debates take place within conferences and workshops; many are reported on the web. The Congressional Record, with its copious add-ins, is published daily. Most congressional hearings on defense matters come into the public domain within weeks. Thus, when there is a debate over, say, whether the U.S. Army should be organized at the division level or the brigade level, the positions get a fairly public airing.[24] Once a decision has been made and the implementation announced (or becomes visible), analysts have plenty of information—

---

[24] See, as an example of one side of the debate, Doug MacGregor, *Breaking the Phalanx: A New Design for Landpower in the 21st Century*, Westport, Conn.: Praeger, 1997.

perhaps a glut of it—explaining why. Perhaps the details of U.S. force composition are too subtle and minor to merit the kind of press that the larger decisions do, but then making too big a deal over what they mean is misguided. Thus, whatever GFM DI information says to foreign analysts about U.S. doctrine that, in turn, influences an enemy's force structure to the detriment of the United States is likely to be so small and indirect as to have as much likelihood of being helpful as it is harmful.

In considering the sorts of planning decisions an adversary would be making based on its understanding of current U.S. force structure and likely future plans, approximation may be good enough. As long as (1) the approximation is only randomly different from the precise; (2) adversaries are not highly risk averse; and (3) adversaries postpone decisions while looking for better data, the decisions likely to be made using approximate information are quite similar to those made using exact information. When it comes to making their own force-structure decisions, the size of the U.S. force structure is only one element, and by no means the most important, in the mix.[25]

All this detail should not be allowed to obscure an overall truth: The size and composition of the U.S. armed forces are globally known (even if some U.S. capabilities for collecting intelligence and pulling off special tricks can only be surmised). This is the case for most global militaries, given the amount of publicly available information.[26] GFM DI information supplies the details, or perhaps the details of the details, on such information. But strategic decisions rarely hinge on such details.

---

[25] Other factors include budgetary constraints, bureaucratic politics, the art of the technologically possible, and equipment-skills mismatch issues. Very few countries these days build their militaries simply to thwart the United States; they face other (and usually closer) threats, and their militaries play many roles (e.g., internal security). Indeed, it is just as likely that other countries are keenly interested in how the United States does things in order to emulate rather than frustrate our practices.

[26] This has not always been so. In the late 1930s, Great Britain consistently overestimated the size of the Luftwaffe and thereby pushed themselves into accommodating Hitler more than the military facts alone would have warranted. See George Quester, *Deterrence Before Hiroshima*, New Brunswick, N.J.: Transaction Books, 1986.

To frame this issue mathematically, estimating the value of GFM DI information in enemy hands requires determining which decisions are likely to be made with more-certain data that would not have been made with the less-certain information or would likely have been made differently with it. Returning to the simple example discussed above, what would an adversary decide differently with an error term of 0.9 in an important factor than it would have decided with an error term of 1.0? To recast the matter quantitatively, assume the adversary believes the likelihood of some information being true is X percent. With confirmation, it is Y percent. Their confirmation is an adverse event if and only if (1) the adversary's decision threshold falls between X percent and Y percent and (2) if the decision is adverse. The closer X and Y are, the less likely the confirmation threshold will fall between the two values. Working through different decisions in which an adversary was "on the fence" about the decision in important ways, more-exact information could have an effect, but the net loss to the United States of adversaries sometimes being able to make a slightly correct rather than slightly incorrect decision lies in the lower decimal places.

Perhaps needless to add, the existence of a difference between X and Y is not a given when people use the word *confirm*. Demand for corroborating information may have psychological, rather than analytical, roots: They decide to act only after they feel better about what they know, regardless of whether what they know had really changed.[27] Just because the effect is psychological, however, does not mean it can be ignored. If adversaries were systematically disinclined to act in the absence of corroboration even if they really did have enough information to act without it, corroborating information persuades people to do what is their interest to do—and thus not in the interest of the United States to do. Thus, releasing it could be harmful. Conversely, if

---

[27] Consider how the portion of the population that was politically involved reacted in polls during the 2008 Presidential election. If information gathering were directly related to the need to make decisions, those who had made up their minds about the candidate would have paid no attention to polls, and those who were undecided would be eager to consume the data. No such finding comes to mind. Rather, people consulted polls, often in a hopeful manner, to confirm to themselves that the future would look like what they thought it would look like.

people are inclined to act when prudence dictates waiting in the face of uncertainty, they may just want corroboration to steel their courage; if so, the effect of useless corroboration could be to confirm them in their uninformed actions. Because acting prematurely works against their interests, it perforce works in ours. Here, releasing corroborating information is good for us. Absent good evidence on whether people are too eager or too cautious, it is difficult to know whether the psychological boost that corroboration provides is a good or bad thing. The entire notion of confirmation touches the ill-patrolled borders that separate information as an aid to decisionmaking, information as entertainment and emotional support, and information as an element in the arsenal of influence and argument. Only the first purpose—as an aid to (the adversary's) decisionmaking—is a significant argument for classification.

Just because an adversary uses GFM DI to confirm most of what they might already know about U.S. force structure does not mean that that there are no potential concerns in this area. For a small subset of the information included in the minimum data set, confirmation could pose some concern. As before, this concern arose regarding information on individual billets—the most detailed "leaves" in the organizational and force-structure "trees" the initiative constructs.

The sensitive confirmation issue relates to the detailed information on billets coupled with the use of an FMID that permits linkages between force-management and personnel data (the latter being outside GFM DI). As discussed previously, one risk would arise if billets were associated with a minimum clearance level required to hold such a billet.[28] The higher the clearance level indicated, the more sensitive the position. Not only does this information reveal the likely working location of highly knowledgeable individuals, it may provide a clue that an otherwise ordinary unit is not so ordinary.[29]

---

[28] The term *would* refers to the fact that clearance level is not a mandatory item within GFM DI, but some org-server maintainers have added it or are considering doing so.

[29] This statement is analogous to what we will say later about military occupational specialty (MOS) codes associated with billets.

A greater risk would arise when billets are matched to individuals. This can occur because the same mandatory identifier (FMID) used in GFM DI is used in personnel databases to indicate which billets an individual now occupies. Personnel databases also keep a history of individuals' former billets. Furthermore, personnel databases point to a large number of other potential pressure points, such as location, family, and certain financial information. Thus, it is feasible for someone to determine which units are sensitive based on their description, identify the more sensitive billets within such units, and then cross-walk over to the personnel databases to determine who held the billet and infer from other personnel information which were the better sources for intelligence recruitment (or easier to threaten or harm if stationed overseas). That noted, value may not be the primary criterion for recruiters from foreign intelligence agencies; potentially more important to the recruiter may be the likelihood that the target can be compromised and, conversely, the odds that an individual, once approached, will report the recruiter's activities (potentially ending his or her career).

To be sure, some hints that a person is a valuable recruit are observable at relatively low risk (e.g., uniform insignia, place of work). RESUMEINT, as previously noted, is also a powerful open-source tool. But GFM DI could help a state adversary confirm a prior assessment that an individual under consideration is sufficiently valuable to recruit. That being so, the decision to somehow restrict such information (or delink billet and personnel data) passes the last three criteria: It provides new information, thereby increasing adversary knowledge, and facilitates decisions of the sort that might harm the United States. Even if there are other ways of gaining similar information, GFM DI merits reexamination to the extent that its information links to outside personnel databases and (1) allows access to very current data and (2) links information on a person and a billet (e.g., required clearance level), thus allowing (3) valid inferences about the person's importance.

Nonstate Adversaries

Because nonstate actors lack the resources to carry out research, they are much less likely to have a good preexisting picture of U.S. forces.

Gaining access to such information has a much larger chance of affecting a nonstate actor's level of knowledge. So, in this case, the data could pass our second criterion for classification.

But what about the third criterion: Would the added knowledge help them make better decisions? Most GFM DI data are simply not relevant to the sorts of decisions that such groups make. Smaller adversaries, those with few resources and constrained opportunities, are rarely interested in force-on-force engagements with the U.S. military. Thus, the overall picture of the U.S. force structure would be largely irrelevant to most decisions such adversaries make. As a rule, nonstate actors operate against people, both civilians and military. While terrorists tend to target civilians because they are softer targets, and the whole point is to gain headlines, military targets (e.g., the USS *Cole*) are also desirable targets. Insurgents tend to target militaries (and other security forces) because they need to demonstrate to their supporters that they are legitimate soldiers (albeit in civilian dress) protecting their populations from oppressors—although civilians may be targeted to demonstrate that the state security forces cannot in fact secure the population.

Such groups may be interested in which forces are arrayed against them at any given time, but smaller opponents are so overmatched that having access to extensive detail on the size and makeup of U.S. forces is beside the point. Whatever interest they have in general technical descriptions of the systems they may have to contend with cannot be satisfied by GFM DI. At best they may be interested in the range of responses they may face from U.S. forces, but whether GFM DI can help them there is unclear.

Could the data GFM DI covers contribute to the more parochial decisions terrorists must make in carrying out their activities? Past attacks nonstate actors have made on military targets have been dominated by bombings of individuals and equipment, and penetrating facilities a few times for attacks inside.[30] Most of the targets were

---

[30] The sources for this analysis include the RAND Terrorism Database, which includes more than 350 attacks on military personnel, vehicles, and installations since 1968. Since it is a database on terrorism, it excludes many attacks on military targets that do not meet

observable or public: installations, large-scale platforms, and notable individuals. Although al Qaeda attacked the USS *Cole*, nothing in a GFM DI–compliant database on the vessel would have added much to the group's operational planning that they could not get more straight-forwardly through surveillance of the vessel in port. To the extent that GFM DI allows such actors to identify small or lightly secured targets (e.g., organizational elements away from major military bases),[31] it might make an attack possible for a group that wants to strike at the U.S. military but that cannot deal with even the standard security measures surrounding most installations or major platforms. For many groups, however, such low-profile targets would be unattractive precisely because they are low profile and thus attacking them has little propaganda value.

Most such groups' operations are also sufficiently approximate that such detail is largely immaterial. For example, the enemy's weapon of choice in Iraq (and, increasingly, Afghanistan) is the improvised explosive device. The intelligence needed for effective employment has more to do with where U.S. forces happen to be going and how they deploy on the road than with any information an organizational chart could supply. This would be true even if the organizational chart were married to a geospatial database. Other weapons of choice have even fewer intelligence requirements (of the sort available from GFM DI) for effective deployment. Suicide bombers, for instance, rely largely on immediate information picked up on the scene, sometimes seconds

---

its definition of terrorist attack. To supplement that analysis, we also examined previous analyses, including a RAND analysis of state small-unit and nonstate attacks on airbases, an examination of force-on-force facility attacks, an examination of nonstate sieges of embassies, and more-recent restricted analyses of nonstate-actor attack operations on various civilian targets. See Alan J. Vick, *Snakes in the Eagle's Nest: A History of Ground Attacks on Air Bases*, Santa Monica, Calif.: RAND Corporation, MR-553-AF, 1995; Christina Meyer, Jennifer Duncan, and Bruce Hoffman, "Force-on-Force Attacks: Their Implications for the Defense of U.S. Nuclear Facilities," Santa Monica, Calif.: RAND Corporation, N-3638-DOE, 1993; and Brian Michael Jenkins, *Embassies Under Siege: A Review of 48 Embassy Takeovers, 1971–1980,* Santa Monica, Calif.: RAND Corporation, R-2651-RC, 1981.

[31] That said, GFM DI has precious little location data and does not discuss security (except very indirectly, in terms of the number of individuals authorized for assignment to security codes).

before detonation. Area weapons (e.g., large-scale bombs or unconventional weapons) do not need precise information for operational planning because variances in area effects relative to specific aimpoints are likely to overwhelm feasible error rates in aimpoints relative to targets.

Some subsets of the information in GFM DI could contribute modestly to planning for some types of operations. In most cases, attackers could obtain such information through other means (e.g., direct surveillance), although at greater personal and operational risk than the alternatives available to state actors would have. One example would be planning for an attack on a guarded facility; the attackers might use functional information about its units and organizations to help identify attractive attack locations and garrison force data to assess the likely guard force. Another example would be planning for a saturation attack, in which attackers seek to stage multiple operations to saturate the defense's ability to respond to them; the data might help attackers determine how many bombings would be enough to overcome the number of explosive ordnance disposal teams on hand or how many chemical, biological, radiological, and nuclear attacks to deplete the specialized resources on hand to remediate them.[32] Finally, the data could inform personal attacks on individuals occupying particular functions or roles for threats, assassination, or kidnapping (these are analogous to how state adversaries use such information to identify potential intelligence assets).[33]

---

[32] See discussion of this tactic as a response to defensive measures in Brian A. Jackson, Peter Chalk, R. Kim Cragin, Bruce Newsome, John V. Parachini, William Rosenau, Erin M. Simpson, Melanie Sisson, and Donald Temple, *Breaching the Fortress Wall: Understanding Terrorist Efforts to Overcome Defensive Technologies*, Santa Monica, Calif.: RAND Corporation, MG-481-DHS, 2007.

[33] For example,

> [b]y contrast, very few stories of war heroes surfaced during the Balkans conflict. Part of the reason was a Pentagon policy on withholding the names of pilots flying missions in the region, implemented in the wake of reports that pilots' families had received threatening e-mail messages. (Tom Shoop, "The Pentagon vs. the Press," *Government Executive*, November 1, 1999)

Will the Potential for GFM DI to Make Adversary Aggregation of U.S. Force-Structure Information Easier Create Security Concerns?

Even if every individual data point in a database is benign, might not some combination of data be revealing? Although GFM DI is not a database as such (the initiative itself does not bring together all data on U.S. force structure but rather builds an infrastructure by which it is easier to do so), the fact that it standardizes practices and makes it easier for DoD users to bring individual elements of information together, to link and combine them, suggests that it would make it easier for adversaries or potential adversaries to do likewise. That noted, if they knew they wanted such data, the question of "easy" would be unimportant.

As noted, the aggregation argument has two parts:

- *macroaggregation*—the concern that the easier aggregation of force-structure information will provide adversaries easier access to the overall view of U.S. force structure
- *microaggregation*—the concern that the easier aggregation of information about units or lower-level force-structure elements may reveal more-specific information about the individual force-structure elements. [34]

Macroaggregation—GFM DI as Providing the Big Picture of U.S. Force Structure

Whether GFM DI would make it easier for adversaries to aggregate data to form a big picture of U.S. force structure takes a previous question but asks it in a more functional way: Can adversaries build that big picture more easily if they can access post–GFM DI databases than they can with existing sources of information? The fundamental coun-

---

[34] For a related discussion, see Arvin S. Quist, "Classification of Compilations of Information," K-25 Site Classification and Information Control Office, Central Safeguards and Security Organization, K/CG-1096, June 1991. This document discusses the distinction between compilations with no value added as a result, those with substantive value added, and those that would save an adversary substantial effort.

ter to the belief that classifying data GFM DI covers will limit the ability of state adversaries to aggregate data they now get from multiple org servers is that such states can already aggregate such data. After all, U.S. analysts regularly collected information from open sources and created solid aggregate pictures of Soviet and Chinese force structures—nations much more closed than the United States is.

This, nevertheless, leaves the argument that GFM DI reduces their cost to put together such data. After all, the same standardization that makes data aggregation easier for U.S. analysts no doubt confers similar benefits to adversary analysts (if they can get the same data), and more sharing of force-structure information may mean that more information can be found in one place. Thus, adversaries would save resources that they could then reallocate to activities contrary to U.S. interests.

Such an argument is plausible but not very interesting because it brings up the fourth criterion: Would these resources saved be reallocated against U.S. interests? Would, say, intelligence analysts and collection assets now spending their time building a picture of U.S. force structure be reassigned to other tasks aimed at ferreting U.S. secrets or, alternatively, be otherwise directed or even demobilized, with the resources saved being used for other purposes?[35] Although a purely rational adversary would do the latter, a variety of human behavior—including the tendency of bureaucracies to continue tasks they have done previously—could reduce how much payoff state adversaries would actually gain. Then, it is an open question what any of those resources would subsequently be used for.

As for nonstate actors, the most important weakness in the argument that GFM DI data should be restricted to deny them an aggregate picture of U.S. force structure is our third criterion—the relevance of that information to decisions they are making. Such data, as noted above, is largely irrelevant to most decisions such actors need to make;

---

[35] Even if GFM DI data standardization saves adversaries effort, it is likely to save DoD a great deal more—if only because its people are likely to use the data far more often than others will. More to the point, the dollar the United States saves is a direct and certain benefit, while the subsequent use of any resources an adversary saves may or may not undermine U.S. interests.

the fact that it would be easier to aggregate is largely irrelevant (not to mention that they probably cannot penetrate NIPRNET to get access to the data anyway).

Microaggregation—GFM DI as Providing Information by Aggregating Data About Individual Units or Platforms

Do linkages of data at more-detailed levels reveal new information about individual force-structure elements? Although having to rely on multiple, inconsistent data sources has real costs for friendly activities (a main rationale for GFM DI in the first place), it also costs adversary intelligence collectors, who must rationalize data conflicts in their analyses. As such, the status quo represents an inadvertent deception operation—admittedly against friend as well as foe—in which force-structure elements may be hidden in the noise of conflicting or incompatible data systems.

As a result, some information, while not classified, could be assumed to be protected to a lesser degree through obscurity in the noisy background of force-structure information.[36] As long as systems conflicted and data were incomplete or imperfect, the status quo essentially could provide a level of security to units, individuals, or platforms without having to pay the full costs of explicitly classifying such sensitive units. That form of security relied on the difficulty potential adversaries would have in connecting the dots to essentially allow such units to hide in plain sight.

---

[36] New methods of organizing information, such as GFM DI, often make it easier to find. A trade-off that weighed the pros and cons of keeping sensitive data unclassified and came down in favor of keeping it unclassified because it was obscured might, if run in a GFM DI environment, conclude the reverse, that it should be classified because it can no longer be hidden in the clutter. Oddly enough, this is an argument for keeping GFM DI in general unclassified. Security professionals generally dislike security by obscurity. The best examples are drawn from cryptography. It is considered more reliable to make an encryption algorithm public and protect the key than to obscure the encryption algorithm. In the latter case, your side cannot test it, and thus poor algorithms survive longer than they should. Those who determine what the algorithm is and find its weaknesses, meanwhile, also have no incentive to report. Relying on obscurity substitutes bad habit for analysis and generally makes people lazy. If the other side finds the information, there will be little evidence. One also has to wonder how obscure such information really is.

Information about unclassified military units, major platforms, and their association with installations is largely public record. Thus, the main elements of GFM DI that are relevant in this case are the linkage of billets with either platforms or units and whether the characteristics of those billets would indicate things about the platforms or units that would not be revealed otherwise. Because we are focused on the chance adversaries will be able to obtain new information from these linkages, this concern clearly passes our second criteria for classification, but to assess whether cases exist that would meet all our criteria, it is necessary to get more specific about the types of units or platforms that are of concern.

The following sections address each of the situations of potential concern in turn. However, it is important to observe that, as with the previous discussion of alias fields, such concerns are largely notional. Whether they are real depends on whether there are, in fact, such force-structure elements currently relying on the obscurity provided by the status quo to provide their security.

Microaggregation and Platforms Hiding in Plain Sight. The security concern with linking individual billets and their characteristics to specific platforms is whether doing so would reveal information about their missions or functions. Everyone knows that the United States has units and equipment that are devoted to intelligence collection. Many of these are obvious on inspection, e.g., the wide-bodied jet aircraft bristling with antennae. However, the precise nature of some intelligence assets may not be so obvious, and there are intelligence assets whose missions are concealed, so it is possible for them to operate in sensitive areas without raising alarms. Some of these intelligence platforms are classified and protected by the relevant security requirements. However, if there are such assets whose existence is not classified as such but that have been protected through security via obscurity, GFM DI will make that security approach more difficult or even untenable.

The reason is that GFM DI contains information that may permit an astute analyst to infer intelligence missions carried out by nonobvious platforms. Each asset is associated with certain crew billets, and each billet is associated with a primary (and secondary) MOS. Thus, an otherwise nondescript aircraft that shows intelligence-trained (e.g., signals intelligence–trained) crew members reveals its mission (see Figure 4.1). Linking billets that feature surveillance or cryptological MOS codes to an otherwise nondescript air platform would reveal information about the aircraft that would not otherwise be obvious.[37] Although some insight into the personnel requirements for a particular billet could be observed from the individuals who hold it (e.g., identifying their rank or insight into their service history), the specific skills an individual has may be difficult to identify by inspection.

If so, the ostensible case for classification meets the four criteria: The information can be hidden by classification; it is new information that advances adversary knowledge; by revealing formerly obscured intelligence assets, it prompts adversaries to take countermeasures when they might otherwise not do so; and the result—less intelligence—

Figure 4.1
Information Revealed by Mapping Billets to Sensitive Platforms

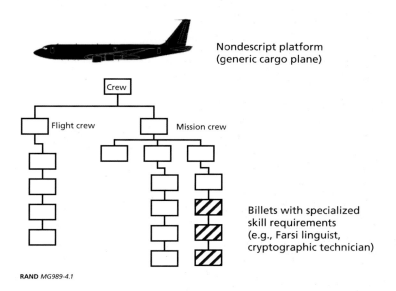

RAND *MG989-4.1*

---

[37] Note that this applies only to billets that are permanent elements of the platform's crew. It does not apply to individuals seconded to the platform or unit for a particular mission (who would maintain their administrative and default relationships to their home organizational elements).

contravenes U.S. interests.[38] Although this notional example is related to intelligence assets, similar arguments could also be made about other sensitive units.

Microaggregation and Units or Individuals Hiding in Plain Sight. Just as the association of individual billets (or individuals themselves) with particular platforms might reveal something about their function, using linkages to determine that data is missing (or that things do not add up) could reveal sensitive force-structure elements otherwise hiding in plain sight. To provide a stable identifier for force-structure elements (including individual billets), GFM DI will assign a stable FMID identifier to each. Because that identifier could potentially be associated with individuals occupying that billet (e.g., through linkage to personnel management databases outside GFM DI), an apparent mismatch between the personal characteristics of the individual and the requirements of the billet (or its associated unit) could tell an adversary that there is something special about that individual that is not otherwise indicated.[39]

Although we have already discussed how such information might be used to confirm that a particular individual held a sensitive position, mismatches in the data—an apparently sensitive position within an otherwise pedestrian unit because its MOS code differs significantly from all others—might limit the ability for individuals with security-sensitive roles (rather than platforms) to hide in plain sight within other units. With legacy systems that were not interconnected and whose data did not always agree, such discrepancies might always be explained as just part of the expected "noise" of systemic errors, and classified entities might theoretically exist in the cracks between systems. Rationalization of data across systems and interlinking the systems make it less straightforward for such protected entities to remain hidden. Furthermore, access to masses of data on the relationship of

---

[38] As noted, this particular system is not the only way units secured only through obscurity might be revealed, but it might allow revelations to occur much more often.

[39] Technically speaking, this concern focuses on *index-based aggregation* because the stable indexing of the data (in this case, the FMIDs) is what enables linkages to be drawn across the larger database.

MOS codes to one another may permit analysts to create a baseline for normal against which it would be easier to detect abnormalities. As in the case for microaggregation leading to identification of sensitive platforms, this situation would also clearly meet our four criteria for justifying classification.

Conclusions

We examined the various concerns regarding GFM DI in light of the four criteria we outlined in Chapter Two to assess the potential benefits of classifying the data in part or in total in response to those concerns.

With respect to the minimum data set and interoperability requirements mandated by GFM DI, the concerns raised about the standardization and broader utilization of force-structure information appeared largely unfounded. In considering an overall picture of U.S. force structure that adversaries might gain, the change from the status quo of many alternative sources of information, and the limited decisionmaking advantage of slightly better data, the concerns generally failed at least one criterion for considering classification. Within the requirements imposed by GFM DI on data providers, the one issue that rose to the level of a *possible* concern is the information that might be revealed about a sensitive unit or platform based on the characteristics of the billets associated with it and the linkage of individuals to billets or units through FMIDs (via personnel databases external to GFM DI), which we termed microaggregation.

Although not required by GFM DI, the data accessible through GFM DI raises other potential concerns. The minimum data set appears largely unproblematic, but data providers do have the flexibility to add additional data and force-structure information to databases that come under GFM DI. Such additional data will be as interoperable and as sharable (although not required to be shared) as those in the minimum data set. Depending on what data are added to the system, sensitive information that adversaries cannot readily gain through other means could be included. The use of alias fields to include additional information about individual force-structure components was the main example of this possibility. Although flexibility in building force-structure databases and customizing them as needs evolve may benefit DoD, the new data they support—whose extent and nature we obviously cannot now predict—could create security concerns.

We turn to our thinking about what to do about these concerns in the final chapter.

CHAPTER FIVE

Concluding Observations and Recommendations

It has been argued that GFM DI, by making it easier to bring together and aggregate U.S. force-structure information, would create opportunities for adversaries to gather intelligence. If this argument is true and consequential, the data the initiative covers should be protected by classification. Having laid out a systematic process for examining the security concerns, determining what the security benefits would be from classification, and incorporating our analysis of the minimum data set, **we found no good reason to classify, as a whole, the data covered by GFM DI.** The potential for adversaries to gain additional overarching information about U.S. force structure as a result of the initiative appears not to be a problem, since all adversaries of concern reasonably have access to comparable data already that can support the types of decisions that are made on the basis of opposing force information.

Yet our examination revealed potential security challenges that did not exist previously. Of the changes that the initiative requires of providers, the greatest security concern is the potential for linkages at the billet level to (1) reveal information about sensitive platforms or (2) make it harder to hide sensitive billets within other force-structure elements. The other potential concerns grew from what providers might add to the database beyond the minimum data set. The latter may provide adversaries with useful information they hitherto lacked.

Although these concerns do suggest the need for some corrective actions, *we still do not believe that they merit classification of significant parts of the data GFM DI covers or major changes in the initiative.* Why? Although we identified *potential* security concerns, how such concerns will actually manifest themselves depends on events yet to happen. Whether alias fields, for instance, become a security problem depends on how data providers use them. Since GFM DI is still being implemented, the opportunity exists to preclude that problem from arising in the first place by educating data providers about it. How great the security benefits would be from restricting information that might reveal sensitive elements currently "hiding in plain sight" are driven not just by how many such units there might be—which could not be assessed in the course of our work—but also by trends in the civilian and commercial use of information technology that tend to make such force-structure elements harder to hide *irrespective of GFM DI's mandates.* The size of the latter problem remains to be determined because most of the U.S. force structure comes in standard units, and most sensitive programs are already classified (and hence not in the unclassified GFM DI to begin with).

Because we lack the basis to speak to the nature, magnitude, or even existence of such concerns, it is very hard to justify a blanket level of restriction that would significantly reduce the benefits associated with GFM DI. The benefits of rationalizing these data inside DoD are substantial. Forgoing these benefits to hedge against security concerns that appear relatively modest and hypothetical would not yield a net benefit. *If* adversaries were able to gain significant new insights into U.S. force structure, and *if* these insights would alter their decisionmaking in important ways, and *if* such decisions would undermine U.S. national security interests, the military would have to reestimate GFM DI's benefits because the balance between risk and reward would be a much closer thing. As it is, the decision is not close.

Although security concerns do exist, most of them are not unique to GFM DI. The literature revealed other open-source efforts that could identify sensitive units or the participation of individuals in classified programs without access to anything as comprehensive as GFM DI. Such examples included only subsets of force structure and individual programs, so they represent only proof-of-concept cases of what can be done. Other unclassified data sets—both on NIPRNET and the Internet—provide parts of the information that could allow an adversary with the ability to crosslink data on individuals to billets and platforms and thereby generate new information. Although it was beyond the scope of our current work to identify all such other data sets and associated analytic strategies that adversaries could use to generate information, the mere existence of such alternatives vitiates the benefits of restricting GFM DI in hopes of denying adversaries some advantage.

Recommendations

What would a prudent course of action be? Although substantial changes in the database are unwarranted, modest efforts to note and address security concerns may be useful. We therefore make the following recommendations:

1. **Exercise reasonable caution over the use of alias fields.** Not knowing what people might someday want to include in such alias fields, we cannot determine which of them are likely to be sensitive. One way to handle this situation is for the Joint Staff or OSD to write responsible-use guidelines for the services to remind them about what data not to include. That said, those who feed service and related org servers must already know not to put information that merits classification on NIPRNET; having the Joint Staff or OSD remind the services to that effect would be redundant and may be annoying. Accordingly, we recommend having someone on or designated by the Joint Staff or OSD periodically scan GFM DI for information that should not be there. At a basic level, service or Joint Staff attention to the number of alias fields added to the database is warranted to scout for potential problems. The greater the reliance on alias fields, the more attention will be required to ensure that new fields are not adding sensitive content to unclassified force-structure databases.

2. **Certain attributes in the data fields may have to be truncated to limit how much sensitive information their expres-**

sion reveals. We noted, for instance, that listing the required security classification of billets may reveal information about individuals that may, in some circumstances, make them targets for recruitment. If such fields are to be permitted, some limits need to be set on the display of sensitive classification billets. For example, the field could indicate only that a billet requires a clearance but not what kind of clearance it requires. Similarly, some MOS codes are inherently sensitive; a unit with an unexpected number of such billets calls attention to itself as having special missions. Perhaps sensitive MOS codes—and a few nonsensitive MOS codes for good measure—should be listed as "other" or something similarly uninteresting.

3. **Give serious consideration to explicitly classifying units, platforms, or activities that now guard their security through obscurity.** Although the type of data integration inherent to GFM DI is indeed a threat to such strategies, other advances in information-gathering and analysis technologies offer similar threats. Manipulating data is getting easier over time, and people routinely disclose information in a myriad of intentional and unintentional ways, making things simpler for an opponent. The half-life of such tactics is short—with or without GFM DI—and prudent planners should anticipate as much and adjust accordingly.

4. **Further study is needed on the mutual effects of GFM DI and personnel databases.** The ability to link a person to a billet and a billet to a unit may reveal a great deal more about the unit than is in GFM DI alone (much less than what is revealed in personnel databases alone). The ability to link persons to each other (by linking person to billet to unit to billet to person) allows potential adversaries room for a large amount of social network analysis. This problem may be more urgent than it appears. Denoting such a crosswalk capability as "forthcoming" refers only to what authorized users cannot do now. Yet as long as the billet FMID in GFM DI is the same as the billet FMID in personnel databases those who can hack into and purloin from both databases need not wait on authorization to carry out

their analysis. However, not having analyzed interlinked personnel databases, we cannot offer a concrete recommendation for how to address this problem.

Broader Considerations

As with any information technology, its responsible use is a shared enterprise among the users, system managers, and designers responsible for different elements of the system. In the course of our study, it became clear that many of the possible difficulties we contemplated might arise from a possible breakdown in the shared responsibility for ensuring a proper balance between utility and security on the system, notably in the creation of alias fields. This means that anyone who can add features using an alias table can set the stage for someone else to inadvertently introduce information that might be more sensitive than the current data in the system apparently are. These users need to understand the importance of thinking through additions to the systems. The question is how to maintain the flexibility of the system, yet decrease the chances of a security issue emerging in the future.

Detecting such breakdowns will likely require ongoing review and evaluation. This is particularly true for a system that will almost inevitably evolve over time, and gaining an understanding of how the data involved are changing can help identify errors and other concerns. These reviews should take place frequently at the outset, when the system is in flux. Later, reviews can be less frequent and can offer all parties a look at the data collectively, to see underlying structural changes. Indeed, an early evaluation of the system as a whole, when populated with data, would provide a vital benchmark that would help the data providers work toward consensus on how the system represents forces. Such reviews would not necessarily need to cover all the data available to the overall system; a representative subset would likely suffice.

This combination of indirect review reinforcing good practice in how individual data providers compile the data they share via GFM DI would appear to be the most appropriate approach to addressing the residual security concerns about the initiative. Given the generally conservative attitude of military organizations toward data sharing in the first place, combining education and periodic assessment seems to be the best course of action as the system is used, working to help users both avoid introducing the problems in the first place and catch the few that might make it through data-provider screening at the service level. Periodic reviews seem especially useful and might occur in the context of simply increasing data providers' awareness of what is and is not likely to be readily known about their basic forces. Reviews also provide an opportunity for other DoD entities to sit in and appraise how well GFM DI does—or does not—interlink data, whether it reveals information about any of their assets, and whether they should seek either to protect the assets rigorously or to adjust planning assumptions to adapt to the potential greater transparency of some elements of force structure.

Sample Attribute Labels

Mandatory Aircraft Type Codes

AIRRW   A machine or device capable of atmospheric flight and
dependent on rotating blades for lift.

FIXWNG   A manned machine or device capable of atmospheric
flight and dependent on wings for lift.

LGTAIR   A manned machine or device capable of atmospheric
flight weighing less than the air it displaces.

NKN   It is not possible to determine which value is most
applicable.

NOS   The appropriate value is not in the set of specified
values.

SPACEM   A manned aircraft capable of operating in the region
beyond the earth's atmosphere

Aircraft Type Categories

Autogyro	Glider
Balloon	Helicopter
Bomber	Not known
Transport	Not otherwise specified
Dirigible	Satellite
Fighter	

Aircraft Type Military Main Purpose

Air to air refuelling
Airborne command post (C2)
Airborne early warning
Airborne early warning and control
Air defense
Air superiority
Antiarmour
Airborne relay
Armed assault
Antisubmarine warfare (carrier based)
Antisubmarine warfare (ASW)
Antisubmarine warfare (MPA)
Antisurface
Attack
Attack/strike
Cargo airlift
Command and control
Communications (C3I)
Search and rescue (Combat)
Drone launch
Electronic countermeasures
Electronic countermeasures (Jammer)
Electronic warfare
Fighter-bomber
Fighter-interceptor
Ground attack
Ground attack reconnaissance
Imagery intelligence gathering
Liaison duties
Maintenance overhaul repair
Medical evacuation
Meteorological
Mine countermeasures
Mine warfare

Maritime patrol
Maritime reconnaissance
Multipurpose
Multisensor
Naval
Naval attack
Not known
Not otherwise specified
Passenger airlift
Patrol
Photo mapping
Radio/radar calibration
Reconnaissance
Reconnaissance, ECM
Reconnaissance, photographic
Reconnaissance, radar
Reconnaissance, visual
Search and rescue
Scout
Signals intelligence gathering
Special operations forces
Special purpose
Storage
Tanker
Target/relay, reconnaissance
Tow target
Utility

Bibliography

Aftergood, Steven, "Navy Urges More Classification by Compilation," Foundation of American Scientists, October 1, 2008. As of May 28, 2010:
http://www.fas.org/blog/secrecy/2008/10/compilation.html

Aftergood, Steven, maintainer, "Obama Administration Documents on Secrecy Policy," website, various dates. As of September 24, 2009:
http://www.fas.org/sgp/obama/index.html

Aid, Matthew M., "All Glory Is Fleeting: SIGINT and the Fight Against Terrorism," *Intelligence and National Security*, Vol. 18, No. 4, 2003, pp. 72–120.

Arkin, William, *Code Names*, Hanover, N.H.: Steerforth Press, 2005.

Assistant Secretary of Defense for Command, Control, Communications, and Intelligence, *Information Security Program*, Washington, D.C., DoD Regulation 5200.1, January 1997.

Baker, John C., Beth E. Lachman, David R. Frelinger, Kevin M. O'Connell, Alexander C. Hou, Michael S. Tseng, David Orletsky, and Charles Yost, *Mapping the Risks: Assessing the Homeland Security Implications of Publicly Available Geospatial Information*, Santa Monica, Calif.: RAND Corporation, MG-142-NGA, 2004. As of May 12, 2010:
http://www.rand.org/pubs/monographs/MG142/

Buchalter, Alice, John Gibbs, and Marieke Lewis, "Laws and Regulations Governing the Protection of Sensitive but Unclassified Information," Washington, D.C.: Library of Congress, Federal Research Division, September 2004.

Cohen, Stephen, *Modern Capitalist Planning: The French Model*, London: Widenfeld and Nicolson, 1969.

Colby, William E., "Intelligence Secrecy and Security in a Free Society," *International Security*, Vol. 1, No. 2, Autumn 1976, pp. 3–14.

Director of National Intelligence and Chief Information Officer, Intelligence Community Technology Governance, "Intelligence Community Classification Guidance Findings and Recommendations Report," January 2008.

Dutch Aviation Society, "United States Air Force & Army (1948–Now)," *Scramble*, undated. As of May 12, 2010:
http://www.scramble.nl/usafbase.htm

Executive Order 12958, "Classified National Security Information," as amended through March 28 2003. As of May 12, 2010:
http://www.archives.gov/isoo/policy-documents/eo-12958-amendment.html

Galison, Peter, "The Art of Transmission: Removing Knowledge," *Critical Inquiry*, No. 31, Autumn 2004. As of June 15, 2010:
http://www.uchicago.edu/research/jnl-crit-inq/features/specialarts.htm

Global Force Management Data Initiative (GFM DI), *Concept of Operations*, Washington, D.C.: Joint Staff Force Management (J-8), April 16, 2007a.

———, *Organizational and Force Structure Construct* (OFSC), Washington, D.C.: Joint Staff, Force Management (J-8), October 12, 2007b.

Hackett, James, Nigel Adderley, Andrew Brookes, Jason Alderwick, Mark Stoker, and Hanna Ucko, "The Military Balance 2007," International Institute for Strategic Studies, 2007.

Information Security Oversight Office, *Report to the President 2004*, Washington, D.C.: National Archives and Records Administration, March 31, 2005. As of May 12, 2010:
http://www.archives.gov/isoo/reports/2004-annual-report.pdf

———, *Report to the President 2008*, Washington, D.C.: National Archives and Records Administration, January 12, 2009a. As of May 12, 2010:
http://www.archives.gov/isoo/reports/2008-annual-report.pdf

———, Cost Report for Fiscal Year 2008, Washington, D.C.: National Archives and Records Administration, May 19, 2009b. As of May 12, 2010:
http://www.archives.gov/isoo/reports/2008-cost-report.pdf

ISOO—*See* Information Security Oversight Office.

J-8/MASO, "Global Force Management Data Initiative (GFM DI): Vulnerability Assessment of Classified Data (VACDAT) Study," briefing, May 30, 2008.

Jackson, Brian A., Peter Chalk, R. Kim Cragin, Bruce Newsome, John V. Parachini, William Rosenau, Erin M. Simpson, Melanie Sisson, and Donald Temple, *Breaching the Fortress Wall: Understanding Terrorist Efforts to Overcome Defensive Technologies*, Santa Monica, Calif: RAND Corporation, MG-481-DHS, 2007. As of May 12, 2010:
http://www.rand.org/pubs/monographs/MG481/

Jenkins, Brian Michael, *Embassies Under Siege: A Review of 48 Embassy Takeovers, 1971–1980*, Santa Monica, Calif: RAND Corporation, R-2651-RC, 1981. As of May 26, 2010:
http://www.rand.org/pubs/reports/R2651/

Jervis, Robert, Richard Ned Lebow, and Janice Gross Stein, *Psychology and Deterrence*, Baltimore, Md.: Johns Hopkins University Press, 1985.

Joint Staff, *Capability Development Document for Global Force Management Data Initiative*, Washington, D.C., August 20, 2007.

Kelley, Sara E., "Features: A Selected Bibliography on 'Sensitive but Unclassified' and Similarly Designated Information Held by the Federal Govt," June 10, 2006. As of June 3, 3010:
http://www.llrx.com/features/sbu.htm

Klein, Gary, *Sources of Power: How People Make Decisions*, Cambridge, Mass.: MIT Press, 1998.

Kopp, Carlo, "Radar Warning Receivers and Defensive Electronic Countermeasures," *Australian Aviation*, September 1988. As of May 28, 2010:
http://www.ausairpower.net/TE-RWR-ECM.html

Lebow, Richard, "Miscalculation in the South Atlantic, the Origins of the Falklands War," in Jervis, Lebow, and Stein, 1985, pp. 103–107.

Lindell, Yehuda, and Benny Pinkas, "Privacy Preserving Data Mining," *Journal of Cryptology*, April 15, 2002, pp. 177–206.

Little, Bertis, Walter L. Johnston, Ashley Lovell, Roderick Rejesus, and Steve Steed, "Collusion in the U.S. Crop Insurance Program: Applied Data Mining," in *Data Mining V: Data Mining, Text Mining and Their Business Applications*, Southampton, UK: Wessex Institute of Technology Press, pp. 291–302.

Loeb, Vernon, "NSA Adviser Says Cyber-Assaults on Pentagon Persist with Few Clues," *Washington Post*, May 7, 2001, p. A02.

MacGregor, Doug, *Breaking the Phalanx: A New Design for Landpower in the 21st Century*, Westport, Conn.: Praeger, 1997.

Madill, Donald L., "Producing Intelligence from Open Sources," *Military Intelligence Professional Bulletin*, Vol. 31, No. 4, October–December 2005, pp. 19–26. As of June 15, 2010:
http://findarticles.com/p/articles/mi_m0IBS/is_4_31/ai_n16419803/

Marquand, Robert, and Ben Arnoldy, "China Emerges as Leader in Cyberwarfare," *Christian Science Monitor*, September 14, 2007.

McLane, Brendan R., "Reporting from the Sandstorm: An Appraisal of Embedding," *Parameters*, Spring 2004, pp. 77–88.

Memorandum for Secretaries of the Military Departments et al., "Protection of Controlled Unclassified Information on DoD Information Systems Connected to the Internet," September 22, 2008. As of May 28, 2010:
http://www.fas.org/sgp/othergov/dod/dod-cui.pdf

Mercado, Stephen C., "Sailing the Sea of OSINT in the Information Age," *Studies in Intelligence*, Vol. 48, No. 3, 2004, pp. 45–55.

Metz, Ed, "Capturing Military Information on the Web and Elsewhere," *Online*, September 1, 2004, pp. 35–39.

Meyer, Christina, Jennifer Duncan, and Bruce Hoffman, "Force-on-Force Attacks: Their Implications for the Defense of U.S. Nuclear Facilities," Santa Monica, Calif.: RAND Corporation, N-3638-DOE, 1993. As of May 26, 2010: http://www.rand.org/pubs/notes/N3638/

Military Periscope, website, undated. As of May 12, 2010: http://www.militaryperiscope.com

Mitra, Sushmita, and Tinku Acharya, *Data Mining: Multimedia, Soft Computing, and Bioinformatics,* New York: John Wiley & Sons Australia, Ltd., 2003.

Moore, Andrew, "Statistical Data Mining Tutorials," website, Pittsburgh: Carnegie Mellon University, undated. As of September 2008: http://www.autonlab.org/tutorials/

Moynihan, Daniel P., *Secrecy: The American Experience*, New Haven, Conn: Yale University Press, 1998.

Multilateral Interoperability Programme, website, 2005–2009. As of May 26, 2010: http://www.mip-site.org/

Mulvenon, James C., and Andrew N. D. Yang, eds., "The People's Liberation Army as Organization: Reference Volume, v1.0," Santa Monica, Calif.: RAND Corporation, CF-182, 2002. As of May 12, 2010: http://www.rand.org/pubs/conf_proceedings/CF182/

Obama, Barack, "Classified Information and Controlled Unclassified Information," memorandum for the heads of executive departments and agencies, May 27, 2009. As of May 12, 2010: http://www.fas.org/sgp/obama/wh052709.html

Paglen, Trevor, *Blank Spots on the Map: The Dark Geography of the Pentagon's Secret World*, New York: E.P. Dutton Adult, 2009.

Palace, Bill, "Data Mining," Los Angeles: University of California, Anderson School of Management, undated. As of September 2008: http://www.anderson.ucla.edu/faculty/jason.frand/teacher/technologies/palace/datamining.htm

Pike, John, "US Forces Order of Battle," GlobalSecurity.org, May 19, 2008. As of May 12, 2010: http://www.globalsecurity.org/military/ops/iraq_orbat.htm

Pozen, David E., "Sum of Its Parts," interview transcript, *On the Media*, New York: National Public Radio, November 25, 2005.

———, "The Mosaic Theory, National Security, and the Freedom of Information Act," *The Yale Law Journal*, December 2005.

Pratt, John W., Howard Raiffa, and Robert Schlaifer, *Introduction to Statistical Decision Theory*, Cambridge, Mass.: MIT Press, 1995.

Quester, George, *Deterrence Before Hiroshima*, New Brunswick, N.J.: Transaction Books, 1986.

Quist, Arvin S., *Security Classification of Information*, Vol. 2: *Principles for Classification of Information*, April 1993. As of May 28, 2010:
http://www.fas.org/sgp/library/quist2/index.html

———, "Classification of Compilations of Information," K-25 Site Classification and Information Control Office, Central Safeguards and Security Organization, K/CG-1096, June 1991. As of May 28, 2010:
http://www.fas.org/sgp/library/compilations.pdf

Shachtman, Noah, "Pentagon Researcher Conjures Warcraft Terror Plot," Wired Blog Network, September 15, 2008. As of October 2008:
http://blog.wired.com/defense/2008/09/world-of-warcra.html#more

Shoop, Tom, "The Pentagon vs. the Press," *Government Executive*, November 1, 1999. As of May 28, 2010:
http://www.govexec.com/features/1199/1199media.htm

Stein, Janice, "Calculation, Miscalculation, and Conventional Deterrence: the View from Jerusalem," in Robert Jervis, Richard Ned Lebow, and Janice Gross Stein, *Psychology and Deterrence*, Baltimore, Md.: Johns Hopkins University Press, 1985.

STRATFOR Global Intelligence, naval update map, web page, April 1, 2009. As of September 29, 2009:
http://www.stratfor.com/analysis/20090401_u_s_naval_update_map_april_1_2009

Svan, Jennifer H., and David Allen, "DOD Bans the Use of Removable, Flash-Type Drives on All Government Computers," *Stars and Stripes* (Mideast ed.), November 21, 2008.

UNESCO, COMEST, "The Precautionary Principle," March 2005. As of May 28, 2010:
http://unesdoc.unesco.org/images/0013/001395/139578e.pdf

U.S. Army Intelligence Center and Fort Huachuca, "Military Intelligence Professional Bulletin," Vol. 31, No. 4, October–December 2005.

U.S. Army, "Table of Organization and Equipment," March 8, 2000. As of May 12, 2010:
http://www.fas.org/man/dod-101/army/unit/toe/

"US Army's Concerns with Protection of Controlled Unclassified Information," August 15, 2008. As of May 28, 2010:
http://www.fas.org/sgp/othergov/dod/dib-cui.pdf

U.S. Department of Defense, Office of the Inspector General, "Controls over the Contractor Common Access Card Life Cycle," Report No. D-2009-005, October 10, 2008. As of May 28, 2010:
http://www.fas.org/irp/agency/dod/ig-cac.pdf

U.S. Department of Defense Operations Security Program Manual, 5205.02-M, November 3, 2008.

U.S. Department of Energy, *Report of the Fundamental Classification Policy Review Group*, Washington, D.C., October 1997. As of May 28, 2010:
http://www.fas.org/sgp/library/repfcprg.html

U.S. Department of the Army, "The Soviet Army: Troops, Organization, and Equipment," FM 100-2-3, June 1991 [supercedes version of July 16, 1984].

———, "Open Source Intelligence," FMI 2-22.9, December 2006 (expired December 2008), pp. 2-5 and 2-6. As of May 28, 2010:
http://www.fas.org/irp/doddir/army/fmi2-22-9.pdf

U.S. House of Representatives, *City on the Hill or Prison on the Bay?* Part III: *Guantanamo—The Role of the FBI: Hearing Before the Subcommittee on International Organizations, Human Rights, and Oversight of the Committee of Foreign Affairs*, Washington, D.C., U.S. Government Printing Office, June 4, 2008.

Vick, Alan J., *Snakes in the Eagle's Nest: A History of Ground Attacks on Air Bases*, Santa Monica, Calif.: RAND Corporation, MR-553-AF, 1995. As of May 26, 2010:
http://www.rand.org/pubs/monograph_reports/MR553/

Wang, Key, Philip Yu, and Sourav Chakraborty, "Bottom-Up Generalization: A Data Mining Solution to Privacy Protection," Fourth IEEE International Conference on Data Mining, 2004.

Wasserbly, Daniel, "Army Cyber Task Force to Manage Growing Industrial Espionage Risk," *Insider*, October 20, 2008. As of May 28, 2010:
http://defensenewsstand.com/insider.asp?issue=10202008sp